The Volatility Playbook

Trading the VIX, Skew, and Implied Vol Without a PhD

Joseph Trek

Published by NobleTrex Press

For permissions and other inquiries, write to:

P.O. Box 3132, Framingham, MA 01701, USA

Contents

Introduction

The most dangerous day in the market is rarely the day of the crash. The crash itself is loud, obvious, and splashed across every television screen in the world. You know the crash is happening because your phone won't stop buzzing. No, the most dangerous day is the one right before it-the day when the sky is blue, the indexes are drifting lazily upward, and the cost of protection is cheaper than a cup of coffee.

It is on that quiet Tuesday afternoon, lulled by a low-volatility regime and the comforting illusion of endless liquidity, that portfolios are built to explode.

Welcome to the paradox of volatility. In traditional finance, we are taught that risk and reward are linearly correlated-that to get more return, you must accept more variance. But the professional trading floor tells a different story. In the trenches of the derivatives market, risk is not a static number on a spreadsheet; it is a living, breathing asset class that can be bought, sold, hedged, and harvested.

Most investors spend their lives trying to predict the direction of price. They ask, "Will Apple go up?" or "Is the S&P 500 overvalued?" They look at charts, read earnings reports, and make a binary bet on direction. It is a hard game. You have to be right about the direction *and* the timing.

Volatility traders play a different game entirely. We don't ask where the price is going; we ask *how* the price is going to move.

We trade the magnitude of the fear, the speed of the panic, and the premium that the market demands for uncertainty. We are the insurers of the financial world, collecting premiums when the weather is calm and paying out when the hurricane hits-provided, of course, that we have built our houses out of brick and not straw.

This book is titled *The Volatility Playbook* because trading volatility is not a science; it is a contact sport. It requires a set of plays, a deep understanding of the playing field, and the discipline to stick to the game plan when the crowd is losing its mind. The subtitle promises that we will do this "Without a PhD," and I intend to keep that promise.

For too long, the world of volatility-the VIX, the skew, the term structure, and the Greeks-has been gatekept by complexity. If you open a standard textbook on options pricing, you are immediately assaulted by stochastic calculus, Brownian motion, and partial differential equations. These are beautiful mathematical concepts, but they obscure a simple truth: Trading is about supply, demand, and edge. You do not need to know how to derive the Black-Scholes equation to understand that when people are terrified, they overpay for insurance. You do not need a doctorate in mathematics to see that the market falls faster than it rises.

You just need to know how to read the machine.

The Hidden Engine of Returns

Why should you care about volatility? Perhaps you are a stock picker, or a passive index investor, or a crypto enthusiast. Why bother with the arcane world of implied volatility and gamma?

Because volatility is the engine that drives the price of everything else.

Consider the events of 2020. Or 2008. Or the "Volmageddon" of February 2018. In each of these instances, the equity markets didn't just go down; the structure of the market itself broke. Liquidity evaporated. Spreads widened. The correlations between assets went to one. The investor who only understood "price" was blindside. The investor who understood "volatility" saw the storm clouds gathering on the horizon long before the rain started.

More importantly, volatility offers something that pure directional trading cannot: a structural edge. In the stock market, the "edge"-the statistical advantage you have over the person on the other side of the trade-is fleeting. Information travels at the speed of light. But in the options market, there is a persistent, structural imbalance. Human beings are biologically wired to fear losses more than they value gains. We are crash-averse. As a result, the market persistently overprices the probability of catastrophe.

This phenomenon creates the **Volatility Risk Premium**. It is the tendency of *Implied Volatility* (what the market pays for options) to be higher than *Realized Volatility* (what the market actually does). This spread is the source of return for volatility sellers, much like the premiums collected by an insurance company. But unlike an insurance company, which has actuaries and decades of data, the market is driven by emotion. It panics. It overreacts. And therein lies the opportunity.

However, this is not a brochure for "easy money." If selling volatility were easy, everyone would be rich. Instead, the graveyard of finance is littered with the corpses of traders who sold options to pick up pennies in front of a steamroller. They didn't understand that the steamroller moves faster than they can run. They didn't understand the "tail risk"-the rare, black swan events that wipe out years of profits in an afternoon.

To trade this asset class, you must respect the machine.

Your Roadmap Through the Chaos

We are going to dismantle the volatility complex piece by piece. We won't start with formulas; we will start with reality.

The first act of this book, **Understanding the Machine**, strips away the abstraction of the "stock market" to reveal the plumbing beneath. You cannot trade volatility effectively if you don't understand how orders are filled, why liquidity disappears when you need it most, and why the "mid-price" on your screen is often a lie. We will look at the harsh mathematics of drawdowns and why avoiding a 50% loss is mathematically more important than finding a 100% gain. We will talk about "fat tails"-the reality that market crashes happen far more often than standard bell curves predict.

Once we understand the terrain, we pick up our tools. In **The Options Toolkit**, we will learn the language of the trade. But we will bypass the academic definitions of the Greeks (Delta,

Gamma, Theta, Vega) in favor of their street definitions. Delta isn't just a partial derivative; it's your hedging ratio. Theta isn't just time decay; it's the rent you pay or collect for holding a position. We will visualize these concepts so that you can look at an option chain and see a map of risk, not just a column of numbers.

With the basics in hand, we arrive at the heart of the book: **Cracking the Code**. This is where we distinguish between the past (Realized Volatility) and the future (Implied Volatility). This distinction is the single most important concept in derivatives trading. We will explore why options are usually expensive, why the market pays a premium for protection, and-crucially-when that premium disappears and it's time to run for the hills.

Then, we tackle the celebrity of the volatility world: **The VIX**. Everyone watches it, the media loves it, and almost everyone misunderstands it. We will demystify what the VIX actually calculates (hint: it's not just "fear"). We will navigate the treacherous waters of VIX futures and the term structure-the curve that dictates whether your trade has a tailwind or a headwind. We will also look at the "Widowmakers"-the volatility Exchange Traded Products (ETPs) that have destroyed more wealth than almost any other financial invention. You will learn why holding a leveraged volatility product long-term is mathematically guaranteed to drive your account to zero, and how to avoid being the sucker at that table.

As we move deeper, we will learn to read the market in 3D. In **Reading the Room**, we introduce the concept of **Skew**. The market doesn't fear all moves equally; it fears the crash more than the rally. This creates a "smirk" in the volatility surface. Learning to read this smirk allows you to see where the market is hiding its anxiety. It allows you to construct trades where the odds are stacked in your favor simply because of how the market prices fear.

But theory is useless without execution. **The Playbook** is exactly what it sounds like: a categorization of strategies for every market regime. We will break down when to sell volatility (harvesting the peace), when to buy volatility (catching the crash), and how to trade the "carry" of the VIX curve. We will discuss specific setups like calendars, diagonals, and gamma scalping-not as abstract diagrams, but as tactical responses to market conditions.

Finally, we address the thing that actually determines your success: **Survival**. Volatility trading is unique because you can be right 99% of the time and lose everything on the 100th trade if your sizing is wrong. We will dedicate a significant portion of this book to risk management, position sizing, and the psychological discipline required to handle the swings. We will discuss how to stress-test your portfolio against the worst days in history so that when the next crash comes, you aren't panic-selling; you're shopping.

A Note on Tone and Truth

I am writing this book as if we are sitting across from each other at a bar, after the close, sketching ideas on a napkin. I want to be honest with you about the risks. There is an industry of charlatans selling "passive income" through options selling, promising 2% monthly returns with "no risk." They are lying to you.

Selling volatility is like picking up coins on a train track. The coins are real, and they are plentiful. But the train is coming. The professional knows the train schedule. The amateur is just looking at the coins.

This book is about learning the schedule.

We will avoid the "ivory tower" academic approach. There will be no proofs. When we use math, it will be arithmetic that you can do on the back of an envelope. If a concept cannot be explained in plain English, it is likely not useful for making money.

However, "simple" does not mean "easy." The concepts here require you to rethink how you view money. You have to stop thinking in terms of "dollars gained" and start thinking in terms of "risk-adjusted exposure." You have to become comfortable with the idea that the market is a probability distribution, not a crystal ball.

We are about to step into the engine room of the financial markets. It is noisy, it is messy, and it is fascinating. Once you see the market through the lens of volatility-once you see the matrix of implied probabilities, skew, and term structure-you will never look at a simple stock chart the same way again. The candlestick chart tells you what happened. The volatility surface tells you what the market *feels* is about to happen.

So, put aside your assumptions about "bull markets" and "bear markets." Those are directional terms for directional tourists. We are looking for something deeper. We are looking for the mispricing of uncertainty itself.

Let's open the playbook.

Chapter 1

Understanding the Machine: Markets and Risk Foundations

Most traders treat the market like a video game where the price on the screen is exactly what they get. In reality, you are walking into a bustling bazaar where the shopkeepers-market makers-set the prices to ensure the odds are always slightly in their favor. Before we can trade complex volatility products, we have to understand the rigging of the casino. This isn't just about vocabulary; it's about survival mechanics. We will uncover why the 'mid-price' is a lie, why losing money hurts mathematically more than making money helps, and why the financial storms that 'shouldn't happen' keep destroying portfolios.

1.1 How Markets Actually Work

A price chart is a magic trick. It turns a messy, continuous negotiation—millions of small decisions, cancellations, tiny panics, and sudden bravado—into a clean little line that looks like *the market* calmly moved from $100 to $101.

That line is a summary, not a place you can shop.

If you want to trade volatility for a living (or even just survive it), you have to unlearn the idea that markets are vending machines. You don't insert *BUY* and receive a clean, fair price. You enter a bazaar where the "price" depends on what you're trying to do, how quickly you need it, and whether anyone is excited (or terrified) at the same moment.

At the center of that bazaar is a deceptively simple object: the *order book*. Think of it as a live list of intentions.

- On one side: people willing to *buy* at specific prices.
- On the other: people willing to *sell* at specific prices.

The best (highest) buy offer is called the *bid*. The best (lowest) sell offer is called the *ask* (or *offer*). The gap between them is the *spread*.

That gap is not a rounding error. It's the market's admission that immediacy costs money.

Here's the part most newcomers never fully internalize: for every trade, there is a buyer and a seller—but they don't have to be two humans who showed up at the same time with matching urgency.

Usually they *don't*.

Most of the time, one side is impatient (they want in *now*), and the other side is patient (they're willing to wait for a better deal). The patient side is providing *liquidity*. The impatient side is consuming it.

When you press a market order, you are paying for urgency. You are saying: "I don't want a fair price. I want *a fill*."

So what is an exchange, really? Not "a place where stocks trade." That's like calling a restaurant "a place where food exists."

An exchange is a rulebook plus a matching engine. It takes in orders, ranks them (by price, then by time, and sometimes by additional rules), and matches buyers and sellers when their prices overlap. It also broadcasts quotes so the rest of the world can see the current best bid and ask.

That's the clean part.

The messy part is that modern markets are fragmented. A single stock or ETF may trade across multiple venues, and liquidity may be displayed, hidden, or partially hidden depending on the order type and venue. Your broker's router is constantly making decisions about where to send your order and how to slice it up.

You don't need to memorize the plumbing on day one. You *do* need to understand the consequence: the "price" you see is a snapshot of the best currently displayed opportunities, not a promise carved into stone.

Now, meet the character you've been paying all along: the market maker.

A market maker is a professional liquidity supplier. Their job is to stand ready to buy when you want to sell, and sell when you want to buy. They post bids and asks, and they adjust those quotes constantly as inventory, news, and volatility shift.

If that sounds generous, zoom in on the business model: they buy slightly cheaper than they sell. That difference is the spread. The spread is their toll for providing immediacy and absorbing risk.

When you trade, you're almost always paying someone for taking the other side when you *need* it.

A useful mental model is a lemonade stand that's willing to do business all day, even when the crowd is indecisive. If the stand bought lemons for the same price it sold them, it would eventually get run over by randomness: spoilage, sudden demand changes, and the occasional stampede.

So the stand posts two prices:

- "I'll buy lemons from you for $1.00."
- "I'll sell lemons to you for $1.05."

That $0.05 isn't greed. It's survival. It covers the risk that right after the stand buys your lemons for $1.00, everyone leaves and the stand is stuck with inventory nobody wants—or worse, that news breaks and lemons are suddenly worth $0.80.

That is what a market maker faces, except the lemons are shares and options, and the news can arrive in microseconds.

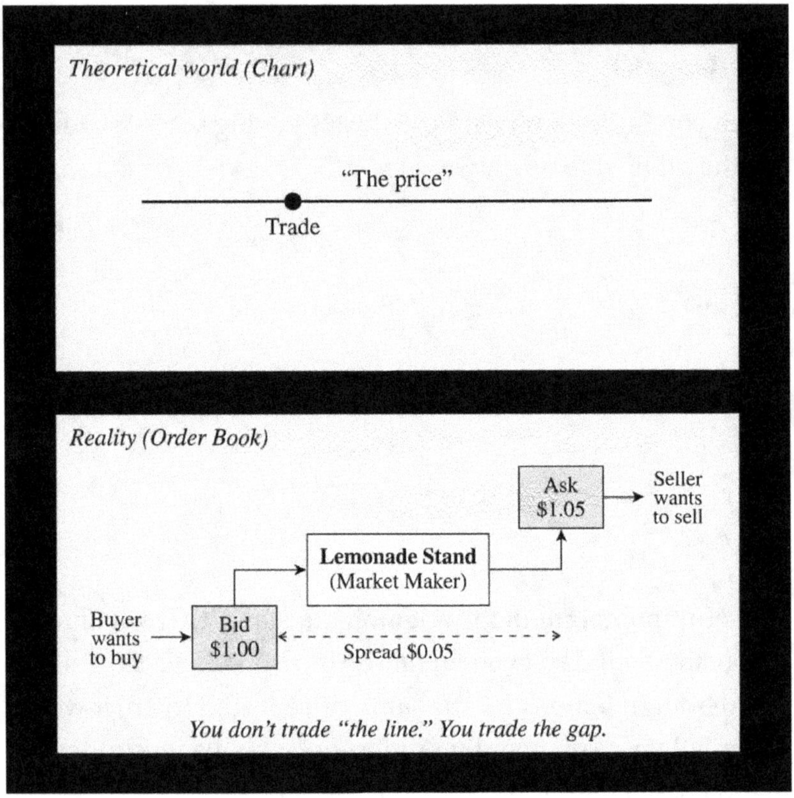

Once you see the spread as a toll, a lot of mysterious trading pain becomes embarrassingly predictable.

You buy, and you're instantly down a little. Not because the market "moved against you," but because you crossed the spread. If the best bid is $1.00 and the best ask is $1.05, and you buy immediately, you pay $1.05. Mark your position honestly a moment later and it's worth roughly $1.00 if you had to sell immediately. That initial dip is the cost of demanding liquidity.

This is why the "mid" price (halfway between bid and ask) is such a seductive liar. It feels fair. It's often what your platform highlights. But you can't hit the mid with a market order. The mid is a *reference point*, not a guarantee.

The order book also has depth, and depth is where fantasy goes to die.

At the top of the book, maybe there are 200 shares bid at $1.00 and 150 shares offered at $1.05. That looks liquid if you only want 10 shares. If you want 2,000 shares, you may not get "$1.05." You may buy 150 at $1.05, then the next 300 at $1.06, then 500 at $1.08, and so on. Your order *walks the book*, chewing through available liquidity at each price level.

That's not manipulation. That's arithmetic plus impatience.

Now add the ingredient that makes volatility trading a different sport: these markets can be thin exactly when you most want them to be thick.

Volatility products—VIX-linked ETPs, VIX futures options, far-out-of-the-money SPX puts, single-name options during earnings—often come with wider spreads and skinnier books than vanilla stocks. Sometimes it's because fewer people trade them. Sometimes it's because the market maker's risk is genuinely nastier: volatility can gap, skew can lurch, and hedging instruments can move in correlated bursts.

When the market maker can't confidently hedge, they defend themselves the only way they can: they widen the spread and reduce size. Translation: the toll road gets more expensive right when the weather turns.

There's also a human layer hiding inside the machine. A lot of liquidity is *conditional*. It's there until it isn't.

In calm markets, market makers compete hard. They narrow spreads because competition forces them to. In fast markets, they pull quotes or back away because getting run over is not a theoretical risk; it's a career-ending one. A market maker who provides tight quotes through a stampede is like a shopkeeper who keeps the doors open while a fire spreads. Noble. Brief.

This is why "liquidity" is not a trait an asset *has*. It's a service other people *offer*, and that service can be repriced in real time.

If you've ever wondered why your fill was worse than expected even though the quote "looked fine," picture the book as a stack of tiny sticky notes.

Each sticky note is a willingness to trade at a specific price for a specific size. Your order doesn't consult a single, authoritative "price." It consumes sticky notes until it's satisfied. If someone cancels a note a millisecond before your order arrives, that liquidity effectively never existed for you.

That sounds unfair until you realize everyone is playing the same game with different equipment and different speed limits.

So where does the last traded price fit into all of this? It's the print that made it onto the tape. It's useful as a historical fact. But treating it as a *current* price is like treating the last car that passed you on the highway as the speed limit. It tells you something about recent conditions, not what you'll get if you hit the gas right now.

This distinction matters in volatility trading because volatility itself is often traded via *derivatives* (options and futures), where the "price" is a premium composed of many moving parts: time, skew, hedging demand, and the market's willingness to warehouse tail risk. In those instruments, the spread isn't merely a nuisance—it's a diagnostic. A suddenly wider spread is the market telling you, in plain language: *I don't feel like catching knives at the moment.*

The market maker is not your enemy. The market maker is a professional whose incentives are simply not aligned with your desire for a perfect fill.

They are paid to:

- provide immediacy *when it is safe enough to do so,*
- manage inventory so they don't get stuck long or short in a panic,
- hedge continuously (often across correlated instruments),
- and earn the spread often enough to survive the times they get clipped.

If you treat them like a villain, you miss the practical lesson: their behavior is information. When spreads widen, size disappears, or quotes flicker, the machine is telling you the cost of liquidity has risen.

And since volatility traders frequently operate in exactly those moments—around macro releases, into the close, during overnight news, on crash days—you are not just trading direction. You are trading the *price of immediacy.*

A good trader learns to ask a better question than "Where is the price going?"

Ask: "What am I paying to be wrong quickly?"

Because spreads are asymmetric pain. Pay them often enough, and you can be right on the big idea and still lose on execution. This is the quiet tragedy of many volatility strategies: the edge is real, the thesis is right, and the P&L leaks out through micro-costs that felt too small to matter.

They matter. They compound.

A final reality check: liquidity is not evenly distributed across time.

Open and close can be jumpy. Lunch hours can be thin. Macro announcements can turn the book into a mirage. Options can widen around earnings. VIX-related instruments can behave differently depending on whether the underlying hedges (like SPX options or futures) are smoothly trading.

The chart doesn't show you any of that. The order book does.

Actionable takeaway: before placing any trade—especially in volatility products—make the spread your first risk metric. If the spread is wide relative to what you expect to make (or if displayed size is tiny relative to your order), you're not "being picky" by waiting; you're refusing a bad toll road. Use limit orders by default, size down when depth is thin, and treat sudden spread widening as a market-wide warning label, not a random inconvenience.

1.2 Orders, Fills, and the Cost of Business

You can be right about volatility and still lose money because you traded like a tourist.

That sounds harsh until you've watched it happen in real time: a trader nails the macro read, catches the direction, even picks a sensible instrument—and then gives the whole edge back one nickel at a time. A wider-than-expected spread here, a fill that prints in three ugly chunks there, a stop that "protected" them by selling the exact low tick. Execution is the unglamorous place where good ideas go to die.

The market doesn't grade your thesis. It grades your *fills*.

A fill is the only truth that matters: the actual price(s) you got, for the actual number of contracts/shares you traded, at the actual time you traded them. Everything else—the quote you saw, the mid you wished for, the last price your app highlighted in bright colors—is atmosphere.

If that feels too cynical, try a simple thought experiment. Imagine you're buying plane tickets. You can read all the travel blogs you want about "typical prices," but when you hit *Purchase*, the only number that matters is the charged amount on your card. Markets work the same way, except the price can change while you're clicking.

Execution has two enemies, and they're cousins: *uncertainty* and *urgency*. The more urgent you are, the more uncertainty you accept. The more certainty you demand, the more you risk not trading at all.

That's the real trade-off behind the most basic decision you'll make: market order or limit order.

A *market order* means, "Fill me now at the best available prices." It's the financial equivalent of hailing a cab in a rainstorm. You're not negotiating. You're buying time.

Market orders are powerful because they maximize the probability you actually get into (or out of) the position. When volatility is exploding and you *must* reduce risk, that certainty can be worth a lot.

But market orders come with a hidden clause written in invisible ink: "Price not guaranteed." In fast markets, the best available price might be very different from the quote you glanced at a heartbeat earlier. If liquidity thins, your market order doesn't politely wait for conditions to improve. It eats what's on the shelf.

That's how people end up buying the top tick in a panic or selling into a hole that only existed for a second.

A *limit order* means, "Fill me only at this price or better." It's the financial equivalent of saying, "I'll buy the ticket if it's under $400. Otherwise, I'm walking."

Limit orders are your primary tool for not getting mugged by the spread and by momentary air pockets. They give you price control. They also turn you into a liquidity provider, at least for that instant: you're offering to trade with someone who's more impatient than you are.

The cost is obvious the first time it stings: you might not get filled. The market can touch your price, tease you, and then run away. Or it can never come close, leaving you with a perfect execution plan and zero position—which is emotionally annoying when you were "right."

That annoyance is the price of discipline.

There's a psychological trap hiding here. Many traders secretly treat a limit order like a moral statement: "I am a careful person, therefore I deserve this price."

The market does not care.

A limit is not an entitlement. It's an offer. If nobody wants to trade with you at that level, you're not being "cheated." You're being ignored. And being ignored is often a gift, because it means you avoided paying up during a moment when the machine was charging extra.

Slippage is what happens when reality refuses to match the story in your head.

Formally, *slippage* is the difference between the price you expected (often the quote or the last trade) and the price you actually got. Practically, it's that sick little feeling when you realize your P&L is down the instant after you hit the button, and you can't even blame the market for moving—it's just the cost of crossing the spread and walking through thin liquidity.

Slippage has multiple parents:

- *Spread cost*: buying at the ask and selling at the bid.
- *Market impact*: your own order moves the price because it consumes available liquidity.
- *Latency and quote changes*: the book updates while your order is in transit.

They look similar on a P&L statement. They feel identical when you're the one paying.

Here's a scene that plays out constantly in volatility products.

You pull up an options chain on something VIX-linked, or on SPX, or on a single name right before earnings. The screen shows a neat quote, something like 1.00 bid / 1.10 ask. The mid is 1.05 and it sits there like a tempting compromise.

You click buy with a market order, expecting something like 1.10.

Instead, you get a fill at 1.18.

Did you get robbed? Not necessarily. You might have consumed the 1.10 offers instantly, then your order kept going and paid 1.12, 1.14, 1.18 as it climbed through the available size. Or the 1.10 offer vanished because the market maker requoted when the underlying ticked. Or the displayed quote was small size and you were bigger than you realized. Your order did exactly what you told it to: *get me filled*.

The market was merely honest about the bill.

Partial fills are the market's way of saying, "Sure, but not all at once."

When the size available at your price is smaller than your order, the matching engine can fill you in pieces. That can happen with market orders (as you walk the book) and with limit orders (as liquidity comes to you). The final result is an average price that may not match any single quote you saw.

Partial fills are not a technical footnote. They change how you manage risk in fast tape. If you think you're "in" but you're only 30% filled, your hedge might be wrong, your exposure might be lopsided, and your next decision might be based on a fantasy position.

If you trade options actively, learn to watch your *filled quantity* the way pilots watch airspeed.

Stop orders deserve a special warning label because they sound safer than they are.

A *stop order* is a trigger: if the price hits your stop level, your order becomes a market order (or, in a stop-limit variant, it becomes a limit order). Stops are often pitched as protection, like guardrails on a mountain road.

But a stop that turns into a market order is protection with a catch: in a fast move, the market can gap past your stop price. You don't get "stopped at" the stop level; you get filled at the best available prices *after* the trigger. In a calm stock, that difference is usually small. In volatility products, it can be cruel.

Stops can still be useful. They just shouldn't be treated as a price guarantee. They are a *get-me-out* instruction, not a contract.

Execution gets more expensive exactly when your emotions get louder.

When volatility spikes, spreads widen and displayed size shrinks. The book becomes jumpy. Quotes flicker. Market makers protect themselves. Everyone tries to do the same thing at once: de-risk, hedge, buy protection. That crowding is not abstract. It shows up as worse fills.

This is why "slippage" isn't bad luck. It's often the market charging surge pricing.

And volatility products are practically built for surge pricing. They're popular during stress, which is the same time they're hardest to trade cleanly.

There's a professional habit that separates traders who survive from traders who "almost had it."

Survivors don't ask, "What's the price?" They ask, *"At what size?"*

A quote without size is a rumor. If the bid is 1.00 for 1 contract and you want 50, that is not a 1.00 bid in any meaningful sense. It's a sample. Your real executable price is the weighted average of the liquidity you'll actually consume.

Platforms often hide this reality behind friendly UI. The market will reintroduce it the moment you trade bigger than your usual.

Limit orders are the default tool for keeping slippage from eating your lunch, but they work best when you use them like a negotiator, not like a statue.

A thoughtful limit order has three virtues:

- It respects the market (it's placed near where trading is actually happening, not at a wish price).
- It respects your edge (it's not so aggressive that you donate spread needlessly).
- It respects time (it has a plan: adjust, cancel, or walk away if conditions change).

That third point is the one retail traders underuse. "Set it and forget it" is fine for a long-term investor placing a buy order below the market. For a volatility trader, a stale limit order can become a little landmine. You might get filled right as conditions deteriorate—not because you found value, but because everyone else is sprinting away from the price you offered.

Time-in-force settings are the quiet knobs that control how dangerous your orders are.

You don't need to become a market microstructure scholar, but you should know what you're asking for:

- *Day* orders expire at the end of the session.
- *GTC* (good-til-canceled) can sit out there like unattended bait.
- *IOC* (immediate-or-cancel) tries to fill whatever it can right now and cancels the rest.
- *FOK* (fill-or-kill) demands all-or-nothing immediately.

For thin options and volatility-linked products, IOC is often the grown-up compromise: it prevents a lingering limit from getting picked off later, while still enforcing a price ceiling.

Here's a practical way to think about execution costs: every trade has a visible fee and an invisible fee.

The visible fee is commission (and maybe regulatory fees). The invisible fee is the spread plus slippage plus impact. In many liquid equities, the invisible fee can be small enough that you can ignore it and still do fine.

In options and volatility products, the invisible fee is often the entire game.

If you're targeting a strategy that expects to make, say, 10–20 cents per contract on average, paying an extra 10 cents on entry because you crossed a wide spread isn't a nuisance. It's a full tax rate.

A simple discipline helps: always estimate your "round trip" cost before you enter.

Round trip means entry plus exit. If the market is 1.00 bid / 1.10 ask, the naive spread cost of buying then later selling (with no price movement) is about 10 cents. If the instrument is thin and you suspect you'll give another few cents to slippage on both sides, your realistic round trip might be 14–18 cents.

Now compare that number to your expected edge. If your expected edge isn't comfortably larger, you're not trading a strategy. You're trading hope.

Volatility traders also face a special execution problem: sometimes you are trading something that is *not* the thing you care about.

You might care about implied volatility (the market's priced-in future wiggle), but you're executing through options whose premiums also depend on time decay, skew, and the underlying's movement. You might care about VIX exposure, but you're trading an ETP whose behavior depends on futures term structure and rolling. The instrument is a proxy, and proxies magnify execution mistakes because you don't get paid directly for what you think you're trading.

That's why clean entries matter so much. When the instrument is already a translation, you can't afford to add sloppy fills as another layer of distortion.

A practical workflow makes execution less emotional:

Before you place the order, decide which of these you are buying:

- *Immediacy* (then accept you'll pay up and size down)
- *Price* (then use limits and accept you might miss)

Trying to buy both at once is how traders end up chasing: they start with a limit, don't get filled, switch to market out of frustration, and end up paying the worst price at the worst moment. The market loves that sequence because it turns your patience into panic.

There is also a humility test that keeps you from overtrading: if you can't get a reasonable fill, you don't have a trade.

Sometimes the smartest execution is no execution.

When spreads are abnormally wide, when quotes are flickering, when the underlying is whipping around, when your platform shows "no bid" on the strike you wanted—that's not the market daring you to be brave. It's the market telling you the toll is high because the road is icy.

Actionable takeaway: make "expected round-trip cost" a required checkpoint before every entry. Look at the bid/ask and size, assume you'll pay at least the spread once (often more in thin options), and write down a quick estimate of your entry+exit friction in dollars per contract. If that friction is a large fraction of your expected profit, either (a) use a limit order with an IOC-style mindset and smaller size, or (b) skip the trade entirely. Execution isn't a detail—it's the business.

1.3 Keeping Score: P&L and Compounding

A trading account doesn't care how smart you felt on Tuesday.

It doesn't give bonus points for a beautiful thesis, a heroic hold, or the fact that your chart annotations looked like something out of a hedge fund pitch deck. Your account keeps score with one blunt instrument: equity. Up, down, or dead.

That bluntness is a feature. It forces a kind of honesty that volatility trading otherwise tries to wriggle out of. Volatility products love drama: big up days, bigger down days, and the constant temptation to confuse *excitement* with *progress*. If you want to last, you need a scoreboard that doesn't get hypnotized.

P&L is that scoreboard. It stands for *profit and loss*, and it comes in two flavors that behave like very different animals: *realized* and *unrealized*.

Realized P&L is cash you've actually locked in. You closed the position (or reduced it) and the result is now part of your account's history. It's no longer a prediction.

Unrealized P&L is what your position would make or lose *if* you closed it right now. It's a mark-to-market estimate, and it can evaporate with insulting speed—especially in options, where spreads widen, liquidity changes character, and volatility itself can reprice overnight.

If realized P&L is money in your pocket, unrealized P&L is money sitting on the table at a busy restaurant. It might still be yours in five minutes. Or someone might bump the table.

This distinction isn't philosophy. It's survival.

Volatility strategies often manufacture gorgeous unrealized gains because implied volatility can deflate quickly after a scare. But those gains can reverse just as quickly when the next headline hits, when the underlying gaps, or when the market decides protection is suddenly fashionable again. Many traders don't blow up because they were consistently wrong. They blow up because they got comfortable with paper profits and treated them like a trust fund.

The market has a particular sense of humor about that.

Now for the cruelest math in finance: the math of recovery.

If you lose 10% of your account, you don't need a 10% gain to get back to even. You need 11.11%.

If you lose 50%, you need 100% to break even.

If you lose 90%, you need 900%.

That's not a motivational poster. That's multiplication. It's the kind of math that turns "I can win it back" into a long, punishing grind—and it turns a single catastrophic drawdown into a career change.

Why is the recovery number so ugly? Because losses shrink the base you're compounding on.

Start with $100.

- Lose 50% → you have $50.

- To get back to \$100, \$50 must double $\rightarrow$ a 100% gain.

The market doesn't owe you symmetry. Percent down and percent up are not mirror images. Down is gravity; up is a ladder you have to climb rung by rung.

This is the part that rewires your brain if you let it: *the first job of a trading strategy is not to make money. It's to avoid getting knocked out.*

Because compounding only works if you're still in the game.

People love to romanticize huge gains—the 10x, the "I turned \$5,000 into \$100,000," the lucky month that gets screenshotted and posted like a trophy. But the serious secret is more boring and more powerful: modest returns compounded for a long time beat heroics that occasionally crater.

Volatility trading tempts you toward heroics because the instruments can deliver them. They can also deliver the crater.

The Mathematics of Recovery

The recovery equation is simple enough to memorize, and doing so changes how you feel about risk:

If you suffer a drawdown of d (as a decimal), the gain required to recover is:

$$\text{Recovery Gain} = \frac{1}{1-d} - 1$$

So for $d = 0.5$:

$$\frac{1}{1-0.5} - 1 = 1 = 100\%$$

That curve is not linear. It bends upward gently at first, then it starts to roar. Past a certain point, recovery stops being a "good month" and becomes a multi-year sentence.

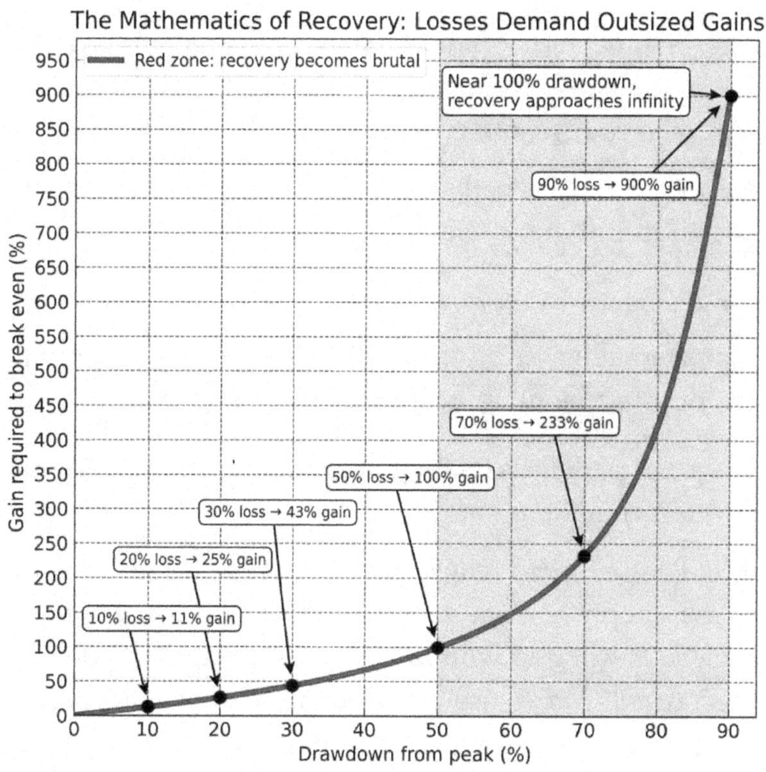

Look at that curve long enough and you'll start to feel suspicious of "aggressive" strategies that brag about average returns while being vague about drawdowns.

That's a healthy suspicion. Average returns are a flattering statistic because they ignore the path. They're like judging a road trip by average speed while ignoring the fact that you drove into a ditch twice.

In trading, the path is the product.

Volatility Tax: When Wiggles Eat Growth

There's another form of cruelty baked into compounding: even if you never experience a catastrophic crash, volatility itself can drag down your wealth growth.

Consider two return paths that both average roughly the same in a simple arithmetic sense:

- A smooth +1% every month.
- Alternating +10%, then -9.09% (which nets to about flat over two months), repeated.

The second path feels exciting. It produces stories. It also produces less compounding.

Why? Because percentages multiply. When you go down, you're multiplying by a smaller number, and the subsequent up move is working with a smaller base. The "wiggles" impose a *volatility tax*—also called volatility drag—that reduces your geometric (compound) growth rate relative to the simple average of returns.

This is why two strategies with the same average return can have very different ending account values. One quietly compounds. The other burns energy turning in circles.

Volatility traders are especially vulnerable to this because many popular approaches are structurally "short convexity" without realizing it: they grind out small wins and occasionally take a large hit. That pattern can look good on a month-by-month hit rate, and it can look even better when the hit hasn't arrived yet.

But the compounding math doesn't care about your win rate. It cares about the size and frequency of the losses that interrupt the multiplication.

Why Realized P&L Deserves Respect (Especially in Options)

In stocks, unrealized P&L can be fairly "sticky" when markets are calm and spreads are tight. In options and volatility products, marks can be slippery.

A quote is not a promise; it's an invitation. If you're long a thin option showing a pretty mid, your unrealized gain may be a mirage if you can only sell at a much worse bid. If volatility drops and the market maker widens the spread, your "profit" can shrink without the underlying doing anything dramatic.

This is one reason professional options traders obsess over execution and over where they can realistically exit. A position that looks profitable on a screen but can't be efficiently unwound is less like an asset and more like a hostage situation.

That doesn't mean you should compulsively take profits the second you see green. It means you should treat unrealized P&L as *exposure*, not as *income*. It is capital at risk until it's realized, and volatility has a habit of repricing your risk faster than your emotions can update.

Keeping Score Like a Pro: Peak, Trough, and the Pain in Between

Most people track returns like a highlight reel: "Up 12% this quarter," "Down 4% this month." That's fine for bragging rights. It's not how you manage a strategy that can gap on a Wednesday.

A trader's nervous system experiences risk as drawdown: how far you are from your most recent peak. Drawdown is the measure that answers the only question your future self will care about during a bad streak: "How deep is this hole, and how hard is it going to be to climb out?"

It's also the measure that forces humility. A strategy can have impressive long-term returns and still be psychologically (and financially) untradable if the drawdowns are deep enough to trigger panic, forced deleveraging, or a simple loss of nerve.

The market doesn't just test your model. It tests your ability to keep placing rational trades while your account is bleeding.

So preserving capital isn't just about avoiding bankruptcy. It's about avoiding the kind of drawdown that makes you change your rules at the worst possible time.

Volatility traders often fail not because their strategy is mathematically doomed, but because the strategy's drawdowns arrive in a form that humans handle badly: sudden, clustered, humiliating. When losses come in a burst, people revenge-trade, double down, or freeze. Any of those reactions can turn a manageable drawdown into a compounding disaster.

The flip side is empowering: small improvements in drawdown control can have outsized effects on long-term wealth. Cutting your worst drawdowns doesn't just reduce pain. It moves you to a gentler part of the recovery curve, where math is still willing to cooperate.

That is the underappreciated edge of risk management: it buys you a friendlier universe.

Actionable takeaway: keep two scoreboards, updated relentlessly: (1) realized P&L (what's locked in), and (2) current drawdown from your account's peak (how deep the hole is). Before taking any new trade or adding size, ask one blunt question: "If this goes wrong, does it push me into a drawdown zone where recovery math becomes punitive?" If the answer is yes, the correct move is usually smaller size, tighter risk limits, or no trade—because compounding only works when you protect the base you're compounding on.

1.4 Defining Risk: Volatility vs. Drawdown

If you ask a room of traders what "risk" means, you'll get two answers that sound compatible until you try to live with them.

The academic answer is volatility: how much returns wiggle around. The trader's answer is drawdown: how far your account falls from its peak before it crawls back. One is a statistic; the other is a stomach feeling with a timestamp.

Both matter. Confusing them is how people build strategies that look brilliant on paper and feel like walking barefoot on broken glass.

Volatility is the market's daily weather. Some places are naturally breezy. Some are calm until they're not. A high-volatility asset has bigger day-to-day moves; a low-volatility asset has smaller ones.

The seductive part is that volatility feels like a complete definition of risk because it's easy to compute and easy to compare. You can rank funds by volatility. You can stick it into spreadsheets. You can annualize it and turn it into a single clean number.

But here's what volatility refuses to tell you: whether those wiggles are deadly.

A strategy can have modest volatility and still blow up. Another can have scary volatility and still compound for years. The difference is not the size of the wiggles. It's what happens when the wiggles line up against you for long enough.

Drawdown is what happens when the market stops being "interesting" and starts being personal.

A drawdown is the decline from your highest account value (your peak) down to a lower point (a trough). Maximum drawdown is simply the worst such peak-to-trough fall over the period you're looking at.

Notice what's hiding in that definition: time and sequence.

Drawdown depends on the *path* of returns, not just the distribution. Two strategies can have the same average return and the same volatility, yet one spends months grinding down in a demoralizing slide while the other bounces around and recovers quickly. Your spreadsheet might call them "equivalent." Your nervous system will not.

A good way to make this concrete is to imagine two traders with the same final annual return.

Trader A has a strategy that pops and drops but generally recovers. Their equity curve looks like a jagged staircase. It's noisy, it's annoying, but it keeps finding new highs.

Trader B has a strategy that looks smooth—until it doesn't. It produces small, steady gains, then takes a deep, slow drawdown that lasts long enough to change their personality.

If you only looked at volatility, Trader B might even look "safer." That's the trap: smoothness can be camouflage. Some of the most dangerous strategies on Earth are quiet right up until the moment they aren't.

This is why volatility is often better described as *variability* than as *danger*.

Volatility measures dispersion. It measures movement. It does not automatically measure harm.

Drawdown measures harm because it forces you to confront the one thing traders hate to talk about honestly: what it feels like to lose, continuously, without a clean reset button.

Now add volatility products to the mix, and the gap between these two notions of risk gets wider.

Options and VIX-related instruments don't just move around. They can *reprice*. Spreads change. Liquidity changes. The "same" trade can behave differently depending on whether the market is calm or stressed. You can also be correct about direction and still lose because implied volatility collapses, because time decay accelerates, or because skew shifts against you.

So day-to-day P&L wiggles (volatility of returns) might look manageable, while the tail behavior—the rare, violent episodes—dominates your drawdown experience.

Volatility tells you the ride is bumpy. Drawdown tells you whether the seatbelt holds.

The Ride vs. the Crash

Think of volatility as the turbulence on a flight. Turbulence can be uncomfortable, and it can rattle your confidence. But a turbulent flight still lands safely most of the time.

Drawdown is what happens when the plane loses altitude and doesn't immediately recover.

That distinction matters because traders routinely confuse discomfort with danger. They avoid strategies that feel noisy even when those strategies have robust risk controls, and they embrace strategies that feel smooth even when those strategies are quietly accumulating fragility.

If you only optimize for comfort, you can accidentally buy a strategy that's engineered to hide risk until it arrives all at once.

The most common example is any approach that earns small premiums regularly by selling insurance-like payoffs. It can produce a pleasingly steady return stream in normal conditions. It can also be structurally exposed to a rare but severe event.

In calm markets, the returns look like skill. In stressed markets, the losses look like physics.

A volatility number, averaged over time, can make that approach look tame—especially if the blow-up hasn't happened in your backtest window. Drawdown doesn't let you hide. Drawdown forces you to ask: "When this goes wrong, how wrong can it go, and how long will it take to recover?"

Why Drawdown Is Path-Dependent (and Why That's the Point)

Path-dependence means the order of outcomes matters. A −2% day followed by a +2% day does not return you to where you started. You end slightly down, because +2% is applied to a smaller base.

That same logic scales up to streaks. A cluster of losses can create a drawdown that's much harder to recover from than a scattered set of losses, even if the average return and volatility end up similar.

And clusters are not rare in markets. Stress tends to come in bunches. Volatility tends to cluster. Correlations tend to jump when you least want them to. Drawdown is the metric that "feels" those realities directly.

Volatility, as a single average number, tends to smooth them out.

This also explains something that surprises people when they first trade actively: a strategy can have a modest daily volatility and still grind you down psychologically if it has serially correlated losses—a lot of small losing days in a row.

Your risk report might say, "Relax, the standard deviation is low." Your mind will say, "Why is this thing *always* red?"

Drawdown speaks that language.

Risk of Ruin: The Only Risk That Ends the Game

There's a harsher concept lurking behind drawdown: risk of ruin.

Ruin doesn't have to mean literally hitting $0. In practice, ruin often means:

- getting margin-called and forcibly liquidated,
- suffering a drawdown so large that you can't recover in a reasonable time,
- or losing enough confidence (or external capital) that you stop trading your plan.

Volatility can be high without implying ruin if position sizing is sane and losses are controlled. Drawdown can be low for a long time and still imply ruin if the strategy contains a hidden cliff.

Ruin is why "preserve capital" isn't a slogan. It's the rule that makes compounding possible.

This is also why professional traders talk about *staying power* more than they talk about brilliance. Staying power is the ability to remain solvent and rational through inevitable bad periods.

Volatility measures how wild the ride is. Staying power is about whether your seatbelt and fuel reserves are adequate.

A More Useful Mental Model for Volatility Traders

If you trade volatility, you're constantly tempted to treat "risk" as a single dial you can turn: more risk for more return, less risk for less return. That's a comforting myth.

In reality, you're dealing with multiple dimensions:

- *Noise risk*: day-to-day variance that can shake you out of good positions.
- *Tail risk*: rare events that can dominate your long-term results.
- *Liquidity risk*: the cost of exits rising exactly when you need them.
- *Behavioral risk*: the chance you abandon your rules mid-drawdown.

Volatility mostly talks about the first one. Drawdown is where the others show up to introduce themselves.

This is why some "low vol" strategies are not low risk at all. They often reduce apparent volatility by selling convexity (selling protection), smoothing returns during normal times and concentrating damage in abnormal times.

The result is a return stream that feels stable until the day stability is precisely what you don't get.

Drawdown is the metric that catches this sleight of hand, because it measures the damage when abnormal times arrive.

How to Use Volatility Without Being Fooled by It

Volatility is still useful. It just needs the right job description.

Use volatility to answer questions like:

- "How much can this position swing on a typical day?"
- "How big should my position be so a normal bad day doesn't break my rules?"

Don't use volatility alone to answer:

- "How bad can this get?"
- "Will I survive the worst month?"

Those are drawdown questions. They require thinking in scenarios, stress periods, and path behavior. They require respect for the fact that markets don't distribute pain evenly.

There's a practical benefit to treating volatility and drawdown as separate: it improves your decision-making during stress.

When you see your P&L whipping around, it's easy to interpret the motion itself as danger and react impulsively. But sometimes the right response to high volatility is simply smaller size and better execution—not abandoning the strategy.

Conversely, when the equity curve looks calm, it's easy to get complacent and scale up. Drawdown thinking makes you ask, "Calm according to what? What happens in the stress regime?"

That question has saved more accounts than any fancy indicator.

The Quiet Killer: When Low Volatility Enables Overconfidence

Here's a pattern worth recognizing early: low realized volatility in your P&L often increases risk, because it invites you to lever up.

If a strategy produces tiny daily fluctuations, many traders conclude it's safe and increase size to make the returns feel "worth it." That scaling can work for months. Then the market shifts regimes, the strategy experiences a drawdown it has not rehearsed, and the oversized position turns a manageable loss into an existential one.

The tragedy is that the low-vol period didn't just precede the drawdown. It helped *cause* it by encouraging leverage.

Drawdown-aware traders treat calm periods as a time to build resilience, not to get cocky.

Actionable takeaway: stop using a single number to summarize "risk." For every strategy or position, track (1) volatility of returns to size the trade so normal wiggles don't force bad decisions, and (2) drawdown from peak to ensure the strategy doesn't contain a hidden cliff. Before increasing size, require a drawdown-based rule (for example: no scaling up while you're below your equity peak, and mandatory size reduction after a defined drawdown). Volatility tells you how bumpy today might be; drawdown tells you whether you can keep playing tomorrow.

1.5 When Markets Crash: Fat Tails and Skew

The most dangerous sentence in finance is: "That move was impossible."

Markets have a way of hearing that sentence as a dare.

Most people's intuition about risk is secretly built on a bell curve—the comforting idea that returns are like heights: most outcomes cluster around average, and the extremes are so rare they're practically folklore. In that world, a 5-sigma day is a once-in-a-civilization event. In the actual market, it's a thing that shows up often enough to have nicknames, documentaries, and entire risk departments built around it.

If you trade volatility, you don't get to live in the bell-curve world. You trade the world that keeps breaking it.

A bell curve (a normal distribution) has two seductive properties. First, it's mathematically tidy: if you know the mean and the standard deviation, you can estimate how often big moves should happen. Second, it makes the future feel legible. It turns uncertainty into a spreadsheet where you can assign neat probabilities to ugly outcomes.

And then reality arrives.

Financial markets produce *fat tails*: extreme outcomes happen far more often than the bell curve predicts. They also produce *skew*: the bad outcomes tend to be more violent than the good ones. Stocks often grind upward and then fall like they discovered gravity midair. That asymmetry is not a poetic metaphor; it's a structural feature of how leverage, liquidity, and human behavior interact.

Here's the practical translation: you can run a strategy for months collecting small wins that feel statistically "safe," and then lose several months (or years) of gains in a handful of sessions.

That pattern isn't an anomaly. It's the signature.

If that sounds like doom, good. A little doom is clarifying. The market is not out to get you, but it also isn't obligated to fit your model. When you accept fat tails and skew as normal, you stop being surprised by the only thing that actually hurts: the tail event that arrives on schedule, right after you decided it couldn't.

The Bell Curve Myth (and Why It Keeps Coming Back)

The bell curve myth survives because it's useful in calm times. When markets are stable, daily returns can *look* roughly normal. Your P&L wiggles. Your volatility number behaves. Your risk report feels like it's describing the world.

Then something breaks—a policy shock, a liquidity vacuum, a cascade of forced selling, a correlation spike—and the distribution changes shape while you're still using yesterday's ruler.

The bell curve doesn't fail gradually. It fails like a chair that looks fine until the moment it collapses under weight.

A deeper reason it keeps coming back: humans love simple narratives, and the bell curve is a narrative masquerading as math. It tells a comforting story: "Most days are ordinary. Rare days are almost impossible. If you manage ordinary days, rare days won't matter."

In markets, rare days *are* the story. They set prices, determine survivability, and dominate long-term performance for many strategies. Ordinary days are where you earn the right to still exist when the rare day arrives.

Non-Normality: The Market Has Memory

One reason fat tails happen is that markets have memory. Volatility clusters. Quiet tends to be followed by quiet; stress tends to be followed by stress. If returns were truly independent coin flips, you wouldn't see this kind of weather system behavior.

But markets are not coin flips. They're networks of balance sheets, margin requirements, human psychology, and feedback loops.

When things go wrong, they often go wrong together:

- a selloff triggers volatility targeting and risk-parity de-risking,
- which pressures liquidity,
- which widens spreads,
- which worsens marks,
- which triggers more de-risking.

That chain reaction is how you get a fat tail: not a single surprise, but a system that amplifies surprises.

Another ingredient is leverage. Leverage turns "I'm down" into "I'm forced to sell." Forced selling is a different animal than discretionary selling. It doesn't wait for better prices. It hits bids, and when bids vanish, it finds the next bids down. That's how markets fall faster than they rise: the selling isn't just emotional, it's mechanical.

This is also why execution and liquidity matter so much in volatility products. In stress, the cost of immediacy spikes just as everyone needs immediacy.

A Picture Worth a Margin Call

When you overlay the bell curve with a more realistic "market-shaped" distribution, the difference is not subtle. The center can look similar. The tails are where the lies live.

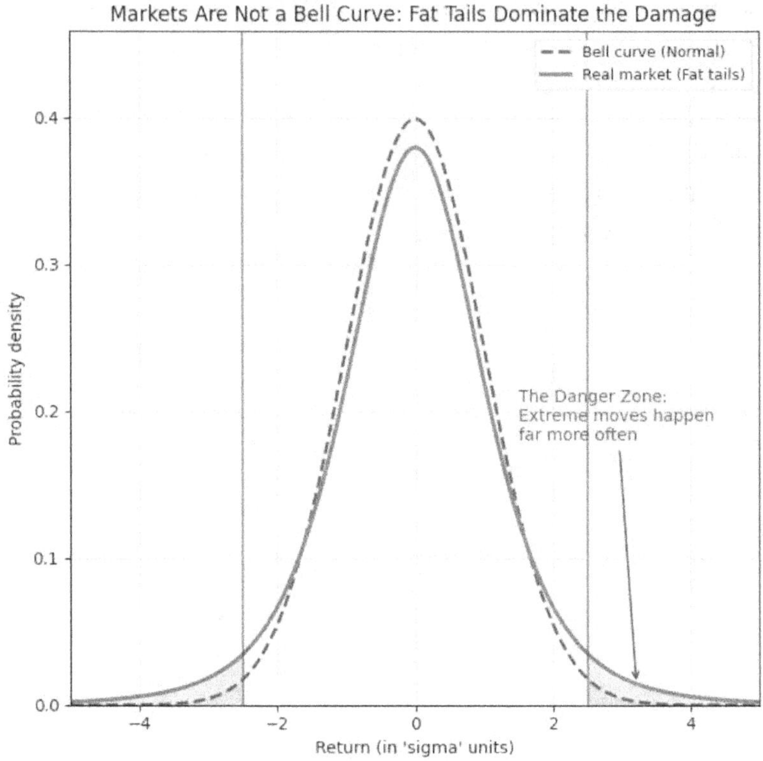

The practical lesson in that red region is brutal: if your strategy's survival depends on those tail moves being "rare," your strategy is built on a fairytale.

This is why certain approaches that look like free money in a backtest are actually time bombs. They are effectively short the tails. They earn small premiums by betting the red zone won't show up often. But the red zone does show up, and when it does, it doesn't politely arrive at the expected frequency.

Skew: Markets Fall Down the Stairs and Climb Up the Elevator (No, Really)

Skew is the market's asymmetry. In equities, downside moves tend to be sharper than upside moves. Part of that is psychology: fear is faster than greed. But a big part is plumbing: deleveraging, margin calls, liquidity withdrawal, and the fact that many investors can't or won't "buy the dip" with the same force that they can be forced to sell.

You see skew directly in option prices. Out-of-the-money puts (downside protection) are typically more expensive, in implied volatility terms, than out-of-the-money calls. The market charges more for crash insurance than for lottery tickets.

That pricing is not a moral judgment. It's supply and demand plus scar tissue.

A good way to feel skew is to remember what a true equity crash does to everyone's calendar. Meetings get canceled. Risk committees convene. Brokers call. People who never talk about markets suddenly ask you if things are "okay."

Up moves don't create that same synchronized urgency. A slow rally does not force anyone to do anything immediately. A fast crash forces everyone to do something *right now*.

Skew is urgency made visible in prices.

Why This Matters for Volatility Trading: Pennies, Steamrollers, and the Insurance Business

Shorting volatility often feels like getting paid for being right most days. You sell options or run strategies that benefit from implied volatility being "too high" relative to what actually happens. In many regimes, that works. You can collect premium, watch it decay, and feel like you discovered a cheat code.

Then the tail event arrives and reminds you what the premium was for.

The premium is not a gift. It is compensation for being the insurer in a world where catastrophes happen more often than a bell curve would suggest.

Insurance businesses can be wonderful businesses. They can also be wiped out by one misunderstood tail.

This is where the earlier ideas about keeping score become sharp. A volatility-selling strategy can show a smooth equity curve (low apparent volatility) for a long time. That smoothness can invite leverage. Leverage magnifies the eventual drawdown. The drawdown moves you up the recovery curve. Suddenly the strategy that "wins all the time" becomes the strategy that takes years to recover—if it survives at all.

The villain isn't volatility. The villain is pretending tails don't exist.

VIX: A Fear Gauge, Not a Crystal Ball

The VIX is often treated like a prophecy: "VIX is at 30, so the market will be volatile." Or worse: "VIX is high, so it has to come down."

VIX is better understood as a price tag. It is an option-implied measure of expected 30-day variance—annualized—

embedded in a strip of index option prices. It reflects not just the market's expectation of future wiggles, but also what people are willing to pay for protection, and what sellers demand to warehouse that risk.

That's why VIX can be "too high" relative to realized volatility on average: the market often overpays for insurance because the emotional utility of protection is high when you remember what crashes feel like.

Treating VIX as a literal forecast misses the economics. It's not weather reporting. It's insurance pricing.

This also explains why "mean reversion" in VIX is not a free lunch. Yes, VIX often comes down after spikes. It also spikes for reasons, and those reasons can get worse before they get better. Selling volatility simply because it looks elevated is like selling flood insurance because the river is already high and you assume it will recede tomorrow.

Sometimes it will. Sometimes you just sold insurance five minutes before the storm hits.

A Trader's Version of Fat Tails: The Day Your Model Stops Working

The scariest part about fat tails isn't the magnitude of the move. It's the way your normal tools become unreliable at the exact wrong time.

Correlations change. Liquidity disappears. Spreads widen. Stops slip. The instruments you hedge with start behaving differently. Your P&L becomes less about your idea and more about your ability to execute under stress.

That's why tail events feel like the market "changing the rules." It's not changing the rules; it's revealing the rules you hadn't priced in.

The only rational response is to build strategies that respect this from the start: position sizes that assume bad gaps, risk limits that anticipate clustered losses, and an acceptance that some percentage of your returns will always be payment for bearing tail risk—whether you acknowledge it or not.

If you don't choose your tail exposure deliberately, it will choose you.

Actionable takeaway: stop evaluating strategies by how they behave on normal days. Instead, force every trade idea to answer one question in plain language: "What happens to me in the red zone?" Concretely: estimate your loss in a fast 2–3 sigma move, assume spreads widen and exits get worse, and size the position so that outcome is survivable without changing your rules. Fat tails and skew aren't rare quirks; they're the rent you pay for playing in real markets.

Chapter 2

The Options Toolkit: Mechanics Before Math

Most people treat options like lottery tickets, but professionals treat them like engineering components. If you want to trade volatility, you don't need to derive the Black-Scholes equation by hand, but you do need to understand the machine's wiring. In this chapter, we are going to strip away the academic fluff and look at the raw mechanics: how calls and puts are actually inextricably linked, why time is an asset class, and how to read the dashboard of risk indicators known as 'the Greeks' without needing a PhD in physics.

2.1 The Building Blocks: Calls, Puts, and Payoffs

Imagine you could walk into a car dealership and buy a coupon that says: *Any time in the next 30 days, I can buy that exact car for $30,000, no matter what happens to prices.* If the market price jumps to $36,000, your coupon is suddenly golden. If the price falls to $27,000, you shrug, toss the coupon in the trash, and buy the car cheaper on the open market. That single idea—a *choice* you can use or ignore—is the soul of an option.

A stock is a commitment: you own the thing and you live with the consequences. An option is a bet with a built-in escape hatch: you pay for the *right,* not the obligation, to transact at a pre-set price. That escape hatch is why options feel like leverage—and why they can be either a seatbelt or a chainsaw, depending on whether you *buy* the right or *sell* it to someone else.

There are only two basic contracts:

- A *call* gives its owner the right to *buy* the underlying at a fixed price (the *strike*) before a specified date (expiration).
- A *put* gives its owner the right to *sell* the underlying at the strike before expiration.

Everything else—every spread, hedge, yield play, tail-risk insurance policy, and "I swear it's low-risk" income strategy—is just Lego bricks built from these two pieces.

The subtle twist that changes your entire risk profile is whether you're the *buyer* or the *seller* (often called the *writer*). The buyer pays a premium up front and gets the choice. The seller receives the premium up front and takes on the obligation: if the buyer exercises, the seller must perform. That obligation is where the drama lives.

Right after you internalize that, payoff diagrams become more useful than any formula. Payoff diagrams don't care about your story. They don't care about your conviction, your macro thesis, or your feelings about the CEO. They show one brutally honest thing: *If the underlying finishes here at expiration, my profit or loss is exactly this.*

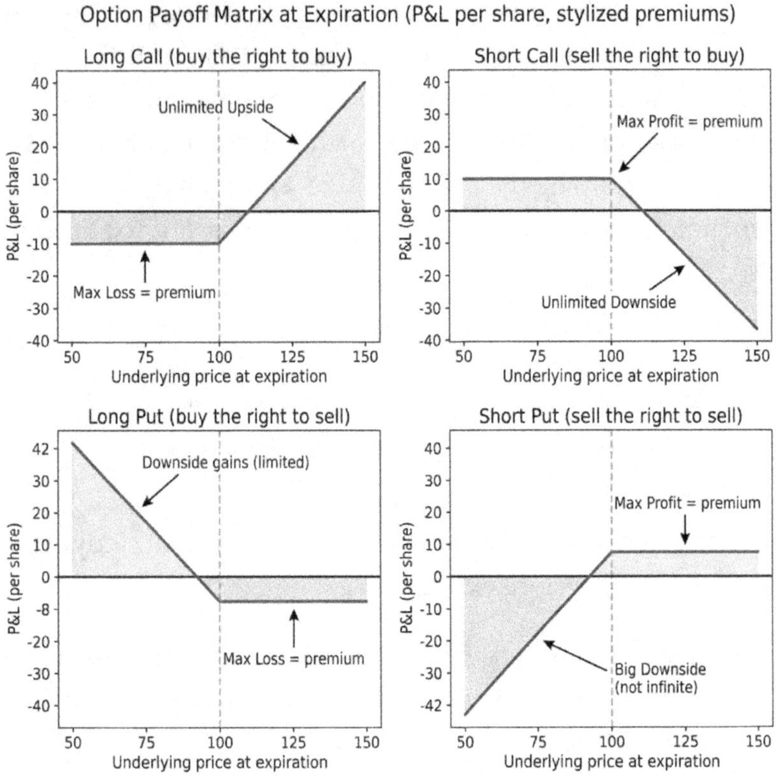

If you stare at these four pictures long enough, two truths start to feel physical.

First: *buying options is buying asymmetry.* The long call and long put have the same personality: the worst case is known in advance (the premium), and the best case can be dramatically larger. That's why options get described as "defined risk" for buyers. You can still lose money consistently, you can still be wrong in humiliating ways, but you generally don't get the kind of loss that makes you call your broker while your hands shake.

Second: *selling options is selling asymmetry.* The short call is the cleanest example. You collect a small, certain premium and accept a loss that can expand without a natural ceiling if the underlying rips higher. The short put is a different flavor: the loss isn't literally infinite (a stock can't fall below zero), but it's still the kind of loss that can dwarf the premium and arrive fast when the market is panicking.

Those two truths are why option trading is less about "bullish" or "bearish" and more about *which shape of pain you can tolerate.* A long option position tends to bleed slowly (time decay) and then occasionally pay you explosively. A short option position tends to pay you slowly (premium) and then occasionally hit you like a dropped piano.

The strike price: the line in the sand everyone pretends is magic

The strike isn't a prediction. It's a contractual boundary. At expiration, it's the fence that separates exercise from non-exercise. If the underlying finishes above the strike, calls matter; below the strike, puts matter. In the payoff diagrams, it's that dashed vertical line: the point where the shape kinks.

If this feels abstract, picture two neighbors arguing over the future price of a house. One says, "I bet it's worth more than $500,000 by June." The other says, "Fine, I'll sell you the right to buy it for $500,000 in June." That $500,000 isn't prophecy; it's the pre-agreed transaction price. Options don't guess the future. They set the terms for a future argument.

Moneyness: ITM, ATM, OTM without the jargon hangover

Traders use a small vocabulary to describe where the underlying is relative to the strike. It sounds technical until you translate it into plain English:

- *In-the-money* (*ITM*): exercising would make sense *right now*.
- *At-the-money* (*ATM*): it's roughly a tie; the underlying is near the strike.
- *Out-of-the-money* (*OTM*): exercising would be pointless *right now*.

For calls:

- ITM if the stock price is above the strike (you'd love to buy at the lower strike).
- OTM if the stock price is below the strike (why buy at the strike when the market is cheaper?).

For puts it flips:

- ITM if the stock price is below the strike (you'd love to sell at the higher strike).
- OTM if the stock price is above the strike (why sell at the strike when the market is higher?).

Moneyness isn't a label for bragging rights. It's a quick way to communicate two things that matter emotionally and financially: *how much of your option's value is "real" right now*, and *how likely the option is to finish with something worth exercising*.

Intrinsic vs. extrinsic: the price is two things wearing one trench coat

An option's price (the premium) is easiest to understand as two components:

- *Intrinsic value*: what you'd get if you could exercise immediately. A call's intrinsic value is how far the stock is above the strike (if at all). A put's intrinsic value is how far the stock is below the strike (if at all).
- *Extrinsic value*: everything else you're paying for. Time remaining. Uncertainty. The market's expectation of future moves. Fear. Hope. All the things that make tomorrow different from today.

Here's the part that changes how you trade: extrinsic value is the *rent* you pay to hold possibility. That rent decays as expiration approaches. So when someone says "options are wasting assets," they're not insulting your intelligence. They're describing the fact that part of what you bought is *time itself*, and time can only move in one direction.

Intrinsic value is simple and stubborn. Extrinsic value is moody and social. It changes when the market gets scared, bored, euphoric, or suddenly attentive. That moodiness is where volatility traders live.

Expiration: when the story stops and the receipt prints

Payoff diagrams are usually drawn "at expiration" because expiration is when the contract stops being poetic. Before expiration, the option price can be influenced by many inputs: time left, implied volatility, interest rates, dividends, supply and demand, and the market's ability to hedge. At expiration, none of that matters. The option is either worth something based on the final underlying price, or it isn't.

That's why these diagrams are such powerful sanity checks. If a trade makes you money only in a narrow, unrealistic slice of outcomes, the diagram will expose it like fluorescent lighting in a casino at 6 a.m.

Long call: the cleanest ``I want upside without marriage'' trade

A long call is the option buyer's classic. You pay a premium for the right to buy at the strike. If the underlying goes nowhere, you lose the premium. If it rallies above the strike, the call starts to gain value at expiration.

The emotional appeal is obvious: you can be wrong in a boring way. Your worst case is a known number. You can size the trade so that the maximum loss doesn't ruin your month.

The hidden cost is also obvious once you accept it: you're paying for a *deadline*. You can be correct about direction and still lose money if the move is too slow, too small, or arrives after the contract expires. Long calls don't reward "eventually." They reward "in time."

Short call: the trade that looks like free money until it looks like a helicopter

Short calls seduce people because the payoff diagram starts above zero. You get paid first. The premium shows up as immediate profit *if nothing bad happens*. That "if" is where accounts go to die.

Selling a call without owning the underlying stock (an uncovered or "naked" short call) can produce losses that keep expanding as the underlying rises. You can be "mostly right" for weeks, collecting small gains, and then one gap up turns months of premium into a single ugly scar.

The diagram tells you the truth in one glance: the profit is capped, the loss isn't. That's not a moral judgment; it's a structural fact. If you ever feel yourself rationalizing this trade with phrases like "it can't go that high," the payoff diagram is your intervention.

Long put: fear with a receipt (and sometimes, insurance with a refund)

A long put is a way to benefit from downside. It's also the purest form of portfolio insurance: the right to sell at the strike even if the market collapses. If you own stock and buy a put, you've effectively installed a floor under your position for the life of the option.

The long put payoff has a satisfying shape: limited loss (premium), meaningful gains as the underlying falls. Unlike the short call, the upside is not literally infinite because the underlying can't go below zero. But for practical purposes, in a fast selloff, long puts can feel like the only calm object in the room.

The catch is that panic is expensive. When the market is already nervous, puts often carry a higher premium because many people want the same insurance at the same time. Buying a put is easy. Buying a put at a good price requires emotional discipline: you have to shop for fire insurance when the kitchen *isn't* already on fire.

Short put: getting paid to promise you'll buy during a panic

Short puts are widely marketed as a friendly income strategy: "Get paid to buy the stock at a discount." That description is not wrong, but it's incomplete in the way a brochure for skydiving might forget to mention gravity.

When you sell a put, you collect premium and agree that if the underlying finishes below the strike (or if you get assigned early in some cases), you may have to buy the underlying at the strike. Your maximum profit is the premium. Your loss grows as the underlying falls, and it can become severe if the underlying drops hard.

There's a reason professionals respect short puts and amateurs romanticize them: the short put is a promise you make *when things are calm* about what you'll do *when things are not*. The moment you most regret the obligation is exactly the moment you're most likely to have it enforced.

``Naked" is not a vibe, it's a risk classification

In options slang, "naked" just means you're not holding an offsetting position that naturally limits the risk. A short call can be "covered" by owning the underlying shares (so you can deliver them if assigned). A short put can be "cash-secured" by reserving enough cash to buy the shares if assigned. Those are operational details, but they change the lived experience of the trade.

Why? Because the market doesn't care about your intent. It cares whether you can meet the obligation when the contract demands it. Being naked doesn't automatically mean you're reckless, but it does mean your risk can expand beyond what your brain intuitively feels when you glance at the small premium collected.

If payoff diagrams are the x-ray, then "naked vs covered" is the difference between seeing a hairline fracture and seeing a clean bone.

A contract multiplier: the detail that turns ``small premium" into real money

Options are typically quoted per share, but a standard equity option contract in the U.S. usually represents 100 shares. That multiplier is the quiet amplifier in the room. A $2.00 option premium is $200 per contract. Ten contracts is $2,000. Fifty contracts is $10,000. The payoff diagram doesn't change, but the emotional consequences do.

A useful habit: whenever you see a premium, translate it into dollars per contract immediately. It forces you to size trades like an adult.

Why payoff thinking beats price-target thinking

Most beginners approach a trade like this: "I think the stock goes to $X." Options punish that mindset because the option is not a stock replacement; it's a *conditional* exposure with a clock attached.

Payoff thinking asks better questions:

- *Where do I make money?*
- *Where do I lose money?*
- *How bad can it get?*
- *What has to happen for this to be meaningfully profitable, not just technically profitable?*

That last question is the grown-up one. Many trades look fine until you realize that your breakeven is far away, your maximum profit is small, or your downside is a cliff that starts exactly where human psychology tends to break.

A payoff diagram also reveals something embarrassingly common: traders often buy options when they secretly want the payoff of *selling* them (steady small wins), and they sell options when they secretly want the payoff of *buying* them (home-run upside). The mismatch between desired emotional experience and actual payoff structure is where bad habits become expensive.

Actionable takeaway

Before placing any single-leg call or put trade, draw the payoff diagram (or at least mentally map its shape) and say out loud:

> *My maximum loss is ___. My maximum profit is ___. My biggest risk is ___. I am being paid/charged ___ to accept that shape.*

If you can't fill in those blanks in 20 seconds, you don't have a trade yet—you have a hunch wearing an options costume.

2.2 Deciphering the Option Chain

The option chain is where good intentions go to die.

Not because options are evil, but because the chain is a glittering spreadsheet that invites the most expensive kind of laziness: seeing a price and assuming it's real. Stocks train you to think that way. Options punish it. In a chain, the number that catches your eye is often a *proposal*, not a *promise*. Treat it like gospel and you'll discover a special kind of pain: the trade that "worked" in theory and still lost money at the fill.

An option chain is simply a menu: for one underlying, it lists calls and puts across strikes and expirations, with quotes and activity data. Most platforms add extra seasoning—implied volatility, Greeks, probability-of-profit doodads, maybe even a little sparkline to make you feel like you're doing something sophisticated. The trick is to learn which fields are nourishment and which are candy.

Start with the contract ID: what are you actually buying?

Every row in the chain is a specific contract, even if the interface makes it feel like a "call at 105" is a universal object. It isn't. Options are defined by four labels:

- Underlying (the stock/ETF/index)
- Expiration date (the deadline)
- Strike price (the deal price)

- Type (call or put)

If you're ever confused, force yourself to read the contract like a sentence: *"I'm looking at a call on XYZ that expires on Friday and gives the right to buy at 105."* That one habit prevents an entire class of errors: trading the wrong expiration, the wrong strike, or the wrong side (call vs put) because your eyes were on the premium instead of the identity.

The chain also tempts you into thinking strikes are evenly spaced and expirations are interchangeable. They aren't. Some expirations are liquid highways; others are dirt roads. Some strikes are crowded; others are ghost towns. Your job is not to find the "best" option. Your job is to find an option you can *enter* and *exit* without donating half your edge to the spread.

Bid, ask, and the ``mid": three numbers, only two are real

Here's the uncomfortable truth: you don't get to trade the pretty number in the middle unless you earn it.

- The *bid* is what someone is currently willing to pay (if you want to sell).
- The *ask* is what someone is currently willing to accept (if you want to buy).
- The *mid* is the average of the two. It's a reference point. It is not a legal obligation from the market.

The mid exists because humans crave a single "price." It makes charts clean, P&L widgets happy, and backtests less annoying. But in live trading, the spread is the toll booth. If you ignore it, you'll keep choosing roads with hidden fees.

A stock with a one-cent spread trains you to be sloppy. An option with a 60-cent spread turns sloppiness into a business model—just not yours. The market maker isn't stealing; they're charging for liquidity, inventory risk, and the privilege of immediacy. Your account doesn't care *why* you paid the toll. It only knows you did.

A simple rule that saves real money: if the spread makes you wince, the trade is already fighting uphill. You're starting the position in a hole, and your thesis has to climb out before you even break even.

The spread is not just a cost; it's a personality test

Two options can have the same mid price and radically different tradability. The difference is the spread and the size behind it.

A tight spread with meaningful size (the number of contracts quoted) says: *"You can probably get in and out without a circus."* A wide spread with tiny size says: *"You're negotiating in a back alley."*

That back alley matters more in options than in stocks because option prices can be small in dollar terms. Paying an extra $0.10 on a $2.00 option is a 5% execution haircut. Pay that on entry and exit and you've handed away roughly 10% before the underlying even twitches. If your strategy expects to grind out small edges, you just vaporized them.

And here's the more subtle pain: spreads widen exactly when you want to trade. Volatility spikes, markets gap, everyone panics, and liquidity becomes expensive. That's not a glitch;

it's the entire reason volatility has a risk premium in the first place. If you plan to be active during stress, you must plan for stress pricing.

Volume vs open interest: ``who showed up today" vs ``who lives here"

Chains usually show two activity columns that sound similar enough to blur together in the brain:

- *Volume* is how many contracts traded *today*.
- *Open interest* is how many contracts currently exist as outstanding positions (not closed, not expired, not exercised).

Volume is the party. Open interest is the neighborhood.

High volume with low open interest can mean a one-day event: traders piling in and out around earnings, a news headline, a re-hedge, a short-term punt. High open interest with low volume can mean a dormant but established position base: a lot of people are parked there, but not many are moving today.

Neither is automatically "good" or "bad," but each tells a different story about your ability to trade without slipping on the spread.

Open interest is often the first filter for avoiding traps. If an option has almost no open interest, it may still trade, but it's more likely you'll see weird, jumpy quotes and fills that feel personal. A chain can look full and still be empty—a grid of strikes where most rows are decorative.

Volume is the second filter for timing. An option with healthy open interest but zero volume today might still be liquid in general, but right now the room is quiet. If you show up with a market order in a quiet room, you become the entertainment.

A useful mental picture: open interest is parked cars, volume is traffic. You want enough parked cars that the place is real, and enough traffic that you can merge without losing a bumper.

The ``ghost town" pattern: how traders get pickpocketed politely

A liquidity trap in options rarely looks like a dramatic warning sign. It looks like opportunity.

You see an option priced at $0.15. Cheap! You imagine buying ten contracts for $150 and turning it into $2,000 if the stock moves. The chain whispers, *"Low cost, high upside."* Then you notice the quote is $0.05 bid / $0.25 ask. The mid is still $0.15, so your brain clings to it like a life raft.

But the trade isn't $0.15. The trade is:

- $0.25 if you buy now,
- $0.05 if you panic-sell now,
- and "somewhere in between" only if you negotiate well and the market doesn't shift under you.

That's the ghost town: low premium, wide spread, low open interest, intermittent volume. It's where lottery-ticket dreams go to be taxed.

The fix isn't complicated. It's emotional. You have to stop equating "cheap" with "good." A cheap option with a nasty spread can be *more expensive* than a pricier option with a tight market, because your round-trip execution cost is the real price of admission.

Bid/ask size: the underappreciated lie detector

Many chains show bid size and ask size: how many contracts are quoted at the best prices. This is not perfect information (quotes can be canceled, hidden liquidity exists, and sizes can be strategic), but it's still a valuable lie detector.

If you see a tight spread but tiny size, that's a fragile peace treaty. If you see decent size on both sides, that's sturdier.

Size also matters when you trade multiple contracts. A beginner mistake is thinking, *"The market is $1.00 wide; I'll just buy 20 contracts."* If the ask size is 1, you may lift one contract at the best ask and the rest at worse prices as you climb the book. Your platform might still show the same quote when you click. Your fill will disagree.

Options don't punish you for being wrong; they punish you for being casual.

Implied volatility and Greeks in the chain: dessert, not dinner

Many platforms display implied volatility (IV) and Greeks right in the chain. They are useful, but they can also become an excuse to overfit a trade: "This strike has higher IV, so it must be overpriced" or "This delta is 0.30, so it has a 30% chance." Those statements can be directionally helpful and still lead to bad decisions if you forget what the chain is actually telling you: *this is where the market is currently quoting risk.*

A practical way to use these fields without getting hypnotized:

- Use IV as a *context marker*. Is the market charging a lot for optionality here, or not much?
- Use delta/gamma/theta/vega as *position labels*, not prophecy. They describe how the option tends to behave *around here*, not what it will do if the underlying jumps, volatility shifts, or time passes.

If you want one piece of IV wisdom that pays rent: IV is often highest where people are most afraid. That doesn't mean it's wrong. It means fear is being priced.

Expiration selection: liquidity often matters more than your cleverness

Chains typically present expirations across the top (or in a dropdown). This is where traders get ambitious and unlucky.

Short-dated expirations are seductive because premiums look "efficient" and deltas feel punchy. Long-dated expirations feel expensive and slow. But execution and liquidity often improve dramatically as you move to the most popular expirations.

Weekly options can be liquid in the front few expirations for major tickers and index products, and completely illiquid for smaller names. Monthly expirations often have deeper open interest because institutions live there. LEAPS (long-dated options) can be wonderfully liquid in large underlyings and a wasteland in others.

The chain is telling you where the crowd is. Trading away from the crowd can be smart when you understand the trade-offs. It can also be a needless self-own.

Strikes: avoid falling in love with round numbers

Round-number strikes (100, 105, 110) often attract open interest. Sometimes that improves liquidity; sometimes it turns the strike into a battleground with strange pinning behavior near expiration. Meanwhile, the "odd" strikes can be either quiet and cheap to trade (nice) or quiet and wide-spread (not nice). You can't know from the strike alone. The chain tells you the truth: spread, size, volume, open interest.

Another quiet strike mistake: choosing a strike because it makes the premium fit your budget. That's backwards. Start with the payoff you want, then pick strikes and expirations that trade well enough to express it.

If your best trade idea lives in a contract you can't enter or exit cleanly, it's not a trade idea. It's a diary entry.

Execution: limit orders are not optional in options

If there's one behavioral upgrade that separates people who get decent fills from people who get farmed, it's this: use limit orders.

A market order in options is essentially telling the market, "*Surprise me.*" The surprise is rarely pleasant. Quotes can move quickly, especially around the open, around news, or in fast markets. The spread can be wide. The book can be thin. A market order is how you turn all of that into a donation.

A practical execution habit:

- Start near the mid (or slightly worse than mid if you want speed),
- wait for a fill,
- and adjust in small increments rather than crossing the entire spread in one impatient click.

The goal isn't to be a hero over a nickel. The goal is to stop paying dimes when you don't have to. Over a year, those dimes become a performance gap you can't explain with your strategy, only with your impatience.

A quick ``chain scan" that actually works

When you pull up a chain, it's easy to get lost in the grid. A simple scan keeps you grounded:

1. Pick the expiration where the bid/ask spreads look consistently sane.
2. Within that expiration, focus on strikes with meaningful open interest.
3. Check today's volume to make sure the room isn't empty.
4. Look at the spread as a percentage of the mid. If it feels like a toll, respect it.
5. Only then look at IV and Greeks as descriptive context.

Notice what's missing: "find the cheapest option" and "find the highest IV." Those are how you optimize for stories instead of fills.

Actionable takeaway

Before entering any options trade, adopt a hard rule: *no chain, no trade*. Open the chain and write down (literally, on paper or in a note) four numbers for the exact contract you intend to trade: bid, ask, open interest, and today's volume. Then answer one question honestly:

> *If I had to exit this position fast, would I trust this market to treat me fairly?*

If the answer is "no," the solution isn't to hope harder. It's to change the contract (different strike/expiration), change the underlying, or change the trade idea. That one discipline does more for real-world profitability than a dozen clever volatility theories that never survive the spread.

2.3 The Iron Laws: Put-Call Parity

If you've ever watched two people argue over the price of something liquid—a concert ticket, a used iPhone, a Yankees seat on a Friday night—you've seen the invisible hand at work. Not as poetry, but as enforcement. Someone posts too cheap, it gets snapped up. Someone posts too expensive, it sits there collecting dust and mockery until the price comes back to earth.

Options markets have their own enforcement mechanism, and it's colder than any comment section: put-call parity. It's the rule that says certain combinations of stock, calls, puts, and cash *must* line up in price, because if they don't, professionals can assemble a near-riskless trade that prints money. In a competitive market, free money doesn't last long. It gets vacuumed up, and the mispricing disappears.

Parity is the reason you should stop thinking of a call and a put as two separate "products" and start thinking of them as two doors into the same room. The doors look different. The room is the same exposure.

A payoff identity disguised as a pricing rule

The cleanest way to understand parity isn't to start with an equation. Start with a promise about the future.

Pick a strike price K and an expiration date. Now imagine two different portfolios you could hold until that expiration:

- Portfolio A: one call at strike K *plus* enough cash set aside to pay K at expiration.
- Portfolio B: one share of stock *plus* one put at strike K.

At expiration, no matter where the stock ends up, both portfolios land in the same place.

If the stock finishes above K:

- Portfolio A exercises the call, uses the cash to buy at K, ends with one share.
- Portfolio B already has one share; the put expires worthless; ends with one share.

If the stock finishes below K:

- Portfolio A lets the call expire, keeps the cash; ends with K in cash.

- Portfolio B exercises the put, sells the share for K; ends with K in cash.

Same ending. Different route.

That payoff identity is parity. Prices may wiggle intraday, spreads may widen, liquidity may hiccup, but the market can't tolerate those two portfolios being meaningfully different in price for long. If Portfolio A is cheaper than Portfolio B, traders buy A and sell B until the discrepancy collapses. If B is cheaper, they do the opposite. That push-and-pull is arbitrage: not a moral crusade, just a business.

The parity equation (without pretending the market is frictionless)

The textbook version for a non-dividend-paying stock is often written as:

$$C - P = S - PV(K)$$

where C is the call price, P the put price, S the stock price, and $PV(K)$ is the present value of the strike (the cash you'd need today to have K at expiration, accounting for interest rates).

If you prefer the "balance sheet" version:

$$C + PV(K) = P + S$$

That reads like a scale: call plus discounted cash equals put plus stock.

In the real world, a few practical frictions blur the edges:

- Stocks can pay dividends, which changes the fair relationship.
- Borrowing stock to short it can be costly or even impossible (hard-to-borrow names).
- Transaction costs and bid/ask spreads turn "free money" into "money after expenses," which is a very different threshold.
- American-style equity options can be exercised early, which complicates the clean European-style derivation.

None of that cancels parity. It just means parity is a *gravity well*, not a glass box. Prices can drift a little, but liquid markets tend to keep the drift bounded because the enforcement is constant.

The first synthetic you should memorize: stock made from options

Parity's most useful gift is that it turns options into a construction set. You can build exposures.

Take the rearranged identity:

$$S \approx C - P + PV(K)$$

That says you can replicate stock (plus financing details) with:

- *Long call*
- *Short put*
- *And cash/financing to account for the strike*

Traders call $+C-P$ at the same strike and expiration a *synthetic long*. It behaves like being long the underlying through that expiration: if the stock rises, the position gains; if it falls, it loses. The payoff shape is essentially the stock's payoff (a straight line), not the curved option lottery-ticket shape.

This is the moment many people realize options aren't "side bets." They are components. You can trade the same underlying exposure through different wrappers, with different capital requirements, different assignment mechanics, and different sensitivity to volatility and rates. That's not theoretical. It's how institutional trading desks think.

Why parity matters to volatility traders (even when you never run an arbitrage)

You might never execute a formal arbitrage package in your life. You can still use parity every week.

Parity explains why options markets feel like a system instead of a collection of independent prices. If calls get bid up aggressively, parity pressure often shows up in puts and/or in the implied forward price. If puts are in heavy demand (hello, fear), the relationship pulls on calls too. The tug happens through market makers hedging and through arbitrage-minded participants who can swap one exposure for another when it's cheaper.

This is also why the option chain can lie to you if you stare at only one side. A call that looks "cheap" might be cheap only relative to a put that's mispriced, or vice versa, or relative to the forward implied by dividends and rates. The market is constantly comparing the two doors into the same room.

Parity is the antidote to naive "I like calls more than puts" thinking. The market doesn't care what you like. It cares what replicates what.

Conversions and reverse conversions: how pros turn parity into a trade

Here's the professional vocabulary, translated into human language.

A *conversion* typically refers to a package that locks in a near-riskless payoff by combining stock and options in a parity-consistent way. One common framing:

- Long stock
- Long put
- Short call

At the same strike and expiration, that package behaves like a fixed payoff at expiration (effectively converting stock into something bond-like for that horizon). Depending on financing, dividends, and borrow costs, it can imply an interest rate. In highly liquid markets, conversions can be used as a financing tool as much as an "arb."

A *reverse conversion* flips the direction:

- Short stock
- Short put
- Long call

Again, the goal is to lock the terminal payoff while harvesting some misalignment in pricing, financing, or borrow.

You don't need to trade these packages to benefit from the idea: they explain *who* steps in when parity is stressed. When mispricings appear, it's not retail traders heroically clicking "buy." It's participants who can short stock, finance cash, and trade multiple legs efficiently. Their activity is why big parity violations are rare and usually temporary in liquid underlyings.

Dividends: the most common reason ``the equation doesn't work"

Equity options are usually written on stocks that may pay dividends. Dividends change the economics because if you own the stock, you may receive dividends; if you own a call, you don't receive dividends unless you exercise and own the stock by the relevant dates.

This is why parity for dividend-paying stocks often uses an adjusted term for the present value of expected dividends over the option's life. In practice, you don't need to model dividends perfectly to understand the main point: dividends tilt the balance between calls and puts because stock ownership has a cashflow that options ownership generally doesn't.

This also points to a very practical, sometimes painful reality: dividend timing is one of the big reasons early exercise and assignment risk exists for American-style options. When dividends are meaningful, the "best" choice for a call holder near expiration can be to exercise early to capture the dividend. If you're short that call, the market doesn't send you a polite calendar invite. It sends assignment.

Parity doesn't just connect prices. It connects behavior.

Borrow and ``hard-to-borrow": when parity is chained to the wall

Another place parity gets interesting is stock borrow. If a stock is hard to borrow (expensive to short, or unavailable), the usual arbitrage mechanisms are weakened. If you can't short the stock easily, some parity trades become impractical, and call/put relationships can look "off" compared to the clean equation.

This matters even if you never short stock, because borrow constraints can show up as weird skews, persistent put richness/cheapness, or strange implied financing in option prices. The market isn't breaking math; it's pricing constraints.

The option chain is not just a set of prices. It's a set of prices embedded in plumbing.

A street-level way to use parity: check whether the market is implying something bizarre

Parity can be used as a sanity check.

If you look at a call and a put with the same strike and expiration, their prices imply an effective forward level for the underlying (loosely: where the market "expects" the stock to be, adjusted for rates/dividends). If that implied forward seems wildly inconsistent with reality, it's either:

- a data issue (stale quotes, wide spreads, bad mid),
- a microstructure issue (illiquid chain, ghost-town strikes),
- a dividend/borrow nuance,

- or a genuine temporary dislocation.

This is where the earlier lesson about bid/ask spreads stops being "execution advice" and becomes "don't get fooled by your own calculator." Parity applied to mid prices in a wide-spread option is like doing physics with a rubber ruler.

When parity looks broken, your first suspicion should be *the quotes*, not *the universe*.

The emotional punchline: calls and puts are siblings, not strangers

Retail culture often treats calls as optimism and puts as pessimism, as if they belong to different religions. Parity says they're family. The market can synthesize one from the other with stock and cash. If you only ever trade one side, you are voluntarily limiting your vocabulary.

More importantly, parity reframes what an option price *means*. It's not just "the market thinks it'll go up." It's the market quoting a set of linked exposures under constraints: financing, dividends, borrow, volatility supply/demand, and hedging flows. Put-call parity is the wire that keeps the whole machine from flying apart.

And for a volatility trader, that wire is sanity. It's what lets you ask better questions than "is this call cheap?": *cheap relative to what, and replicable how?*

Actionable takeaway

Pick one highly liquid underlying you follow (a major index ETF or mega-cap stock). Once a week, do a two-minute parity check using *tradable* prices:

1. Choose a near-the-money strike and an expiration with tight spreads.
2. Write down the call *ask* and put *bid* if you're imagining buying the call and selling the put (synthetic long), or the opposite side if you'd trade it the other way.
3. Compare the call-put difference to the stock level in your head as a rough sanity check. If it looks wildly off, don't conclude "arb"—conclude "something about liquidity/-dividends/borrow is driving this," and investigate before trading.

That habit trains the most valuable instinct in options: never look at a single option price in isolation. In a parity-linked market, every quote has relatives, and the family resemblance is where the truth lives.

2.4 Constructing Spreads and Synthetics

A single option is a strong opinion with sharp edges. You either paid for a lottery ticket (long option) or you sold insurance (short option), and your P&L shape is basically locked in. Spreads are what happens when you stop accepting the default shape.

That's the quiet truth about professional options trading: it's less about being right on direction and more about *engineering a payoff*. You're not asking, "Will the stock go up?" You're asking, "How much up, by when, and what am I willing to pay (or risk) for that bet?" Spreads are the answer to those three questions, expressed in contracts instead of speeches.

There are two big motivations for combining option legs:

- *Define risk*: turn open-ended outcomes into bounded ones so you can size the trade like an adult.
- *Target a variable*: isolate direction, time decay, or volatility so your P&L isn't secretly driven by something you weren't watching.

Most traders learn spreads because their broker platform offers a dropdown. The better reason is that spreads let you decide which dragons you're willing to fight.

The first mental shift: you're trading shapes, not contracts

A long call is a familiar shape: limited loss (premium), unlimited upside. A short put is another: limited profit (premium), ugly downside. Those shapes are real, but they're not destiny.

Once you combine legs, you start sculpting:

- You can *cap* upside to reduce cost.
- You can *cap* downside to reduce existential dread.
- You can *move* the breakeven by paying less premium.
- You can *change* what you're sensitive to: time, volatility, and path.

The market charges for every improvement. That's not cynicism; it's conservation of risk. If you reduce one risk, you usually accept another: capped profit, narrower win zone, greater complexity, or more ways to be "kinda right" and still lose.

Defined-risk doesn't mean ``safe''; it means ``knowable''

"Defined risk" gets marketed like a seatbelt. It's more like a fence around a cliff. The cliff is still there; you just know where it begins.

The most common defined-risk structure is a *vertical spread*: two options with the same expiration but different strikes. You buy one and sell the other. That second leg is not decoration. It's a risk transfer: you're selling away part of your dream to finance the trade and control the shape.

Bull call spread: paying less by selling away your moonshot

Suppose you're bullish. Your first instinct is a long call: buy a call at strike K_1. If the underlying explodes higher, you participate. If it doesn't, you lose the premium.

A bull call spread says: "I'm bullish, but I'm not paying for infinite upside." You:

- Buy a call at a lower strike K_1
- Sell a call at a higher strike K_2
- Both with the same expiration

That short call you sold subsidizes the purchase. In exchange, your profit is capped above K_2. You've traded off the lottery-ticket tail for a cheaper entry and, often, a more achievable breakeven.

This is where payoff diagrams stop being educational and start being operational. The overlay below is the point in one image: the spread has *capped profit*, but it often reaches profitability sooner because you paid less up front.

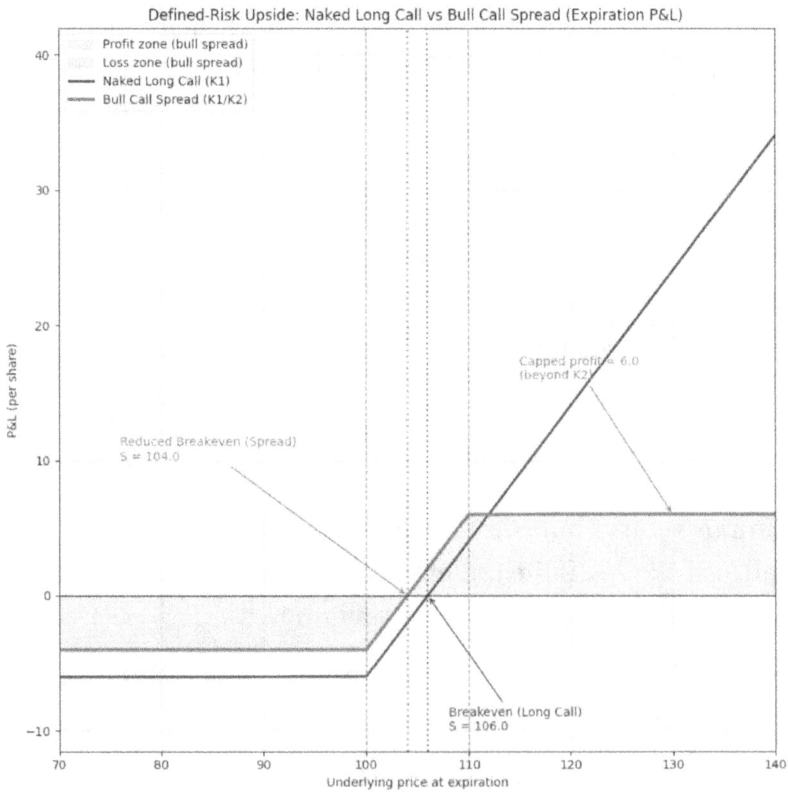

The overlay hides an important psychological trade: the naked call feels heroic because it has "unlimited upside." The spread feels humble because it admits a limit. But markets don't pay you for heroism. They pay you for good bets at good prices. If the underlying needs a gigantic rally for the naked call to work, the "unlimited" part may be irrelevant to your actual life.

One more practical edge: because you sold an option, the spread often has *less sensitivity* to implied volatility changes than a naked long call. That means you're less likely to get whipsawed by volatility deflating after a catalyst. It also means you won't benefit as much if volatility inflates. Every improvement has a bill.

Bear put spread: the same logic, flipped

If you're bearish and don't want to overpay for panic insurance, you can do the mirror image:

- Buy a put at higher strike
- Sell a put at lower strike

The sold put reduces cost but caps your profit once the underlying falls below the lower strike. Again: you're paying less by selling away the extreme tail.

This is the quiet theme of vertical spreads: they're not "beginner strategies." They're *budgeting strategies*. You're telling the market, "I want exposure to a move in a range I consider plausible, and I'm not funding a fairy tale."

Credit spreads: getting paid to be right, and punished for being wrong (in a bounded way)

Debit spreads (like the bull call spread above) require you to pay money up front. Credit spreads do the opposite: you collect premium and accept a defined loss.

A classic example is the *bull put spread*:

- Sell a put at a higher strike (collect premium)
- Buy a put at a lower strike (protection)

If the underlying stays above your short strike, you keep most or all of the credit. If it falls, your loss is capped by the long put you bought.

Credit spreads are popular because they feel like a salary: small, frequent credits. The danger is that they can still be short-volatility trades in disguise. When markets fall fast, spreads widen, and you discover that "defined risk" doesn't mean "gentle risk." It means the maximum is knowable, not that the path to it won't be violent.

Calendar spreads: when time is the thing you're trading

Vertical spreads change the payoff by strike. *Calendar spreads* change the payoff by time.

A simple calendar spread often looks like:

- Sell a near-term option (faster time decay)
- Buy a longer-dated option (slower time decay)
- Same strike

What you're really doing is taking a view on *time decay and implied volatility across expirations*. You're betting the front-month option will lose value faster than the back-month option, and ideally that the underlying doesn't make an ugly directional move that overwhelms the structure.

Calendars are where traders learn humility about "directional" labels. A calendar can be bullish, bearish, or neither depending on where the underlying sits relative to the strike as time passes. The trade's personality can change midstream. That can be a feature (flexibility) or a bug (confusion).

Calendars also teach a market reality: volatility is a *term structure*. The IV for next week and the IV for next quarter are not the same object. When you trade a calendar, you're trading the relationship between them.

Straddles and strangles: buying (or selling) movement itself

If a vertical spread is a controlled directional bet, a straddle is the opposite: it's a bet that the underlying will move a lot, but you're not sure which way.

A *long straddle* is:

- Buy an at-the-money call
- Buy an at-the-money put
- Same expiration

It's an honest trade. You're paying for convexity: big movement pays, stagnation hurts. You're also paying a double premium, which means you need a meaningful move to overcome the cost. Long straddles tend to be long volatility and long gamma—they can benefit from both a big realized move and, often, from implied volatility rising.

A *strangle* is similar but uses out-of-the-money options (cheaper, but needs a larger move to pay).

Selling straddles/strangles is the mirror: you collect premium and bet on quiet. That can feel brilliant until the market stops being quiet. This is where payoff-thinking saves careers: a short straddle's maximum profit is the premium, and the loss can become severe on large moves. If that asymmetry makes you uneasy, listen to the unease. It's your brain reading the payoff diagram correctly.

Synthetics: when options stop being ``options" and start being plumbing

Spreads are payoff engineering. *Synthetics* are identity engineering: using options to replicate something else.

The most famous is the synthetic long stock:

$$\text{Long Call} + \text{Short Put} \approx \text{Long Stock (to expiration)}$$

Same strike, same expiration. Different wrapper.

Why does this matter in practice?

- Sometimes the synthetic is a cleaner way to express exposure than the stock itself (depending on borrow, financing, or portfolio constraints).
- Sometimes it's an arbitrage-ish check: if the synthetic is wildly off from stock economics, something is wrong (or something important is being priced, like borrow or dividends).
- Sometimes it's simply a way to reframe risk: if your "options strategy" accidentally recreates a stock position, you should know that before you congratulate yourself on being sophisticated.

There's also synthetic short stock:

$$\text{Long Put} + \text{Short Call} \approx \text{Short Stock (to expiration)}$$

Again: same exposure, different mechanics.

Synthetics are where put-call parity stops being a concept and becomes a habit. You start seeing trades not as named strategies but as equivalences. The mental payoff is huge: you can sanity-check your own risk. "Am I actually long the underlying here?" "Did I just build a short-vol position without meaning to?" "Is this "income trade" basically a disguised leveraged stock bet?"

Box spreads: the market's ``interest rate'' hiding inside options

A box spread is a combination of a call spread and a put spread with the same strikes and expiration. The payoff at expiration is (nearly) fixed: the difference between the strikes. In a frictionless world, that would behave like lending/borrowing at a risk-free-ish rate implied by options prices.

In the real world, boxes come with footnotes in bold ink: transaction costs, assignment risk (American-style equity options), and financing constraints can turn "clever" into "catastrophic" if you treat it like a free lunch. The important takeaway isn't "go trade box spreads." It's that options markets often embed financing information, and professional players constantly compare options-implied financing to real-world financing.

Once you see that, you stop thinking of options as a casino side room. It's more like a power grid: derivatives, stock, borrow, dividends, and rates all connected.

Why spreads are a volatility trader's home turf

Volatility trading is rarely about one leg because volatility itself is rarely isolated in one instrument. When you buy a single option, you get direction exposure, volatility exposure, and time exposure all tangled together. Spreads are how you start unbraiding that rope.

A vertical spread can reduce how much implied volatility matters relative to direction. A calendar can emphasize the time dimension and term structure. A delta-neutral straddle can lean into volatility and gamma. Synthetics let you separate "I want stock exposure" from "I want option convexity."

In other words: spreads are not a complication layered on top of options. They're how you make options behave like the specific instrument you actually meant to trade.

Actionable takeaway

Before placing any multi-leg trade, do a two-step "shape audit" that takes five minutes and prevents months of confusion:

1. Draw (or mentally sketch) the expiration payoff and label three points: *max loss, max profit, breakevens*. If you can't label them quickly, the structure is still a fog machine.

2. Write one sentence that starts with: *"I am paying (or collecting) premium because I believe..."* and finish it with the variable you're actually trading: *direction in a range*, *time decay*, *volatility expansion/contraction*, or *relative pricing between expirations/strikes*. If the sentence sounds like three variables at once, reduce the position or pick a simpler structure.

The market will happily let you trade a complicated spread for the thrill of feeling advanced. The goal is the opposite: trade the simplest shape that expresses your view, with risk you can name without flinching.

2.5 The Greeks: Measuring Exposure

Options have a reputation for being complicated, but the part that actually hurts people isn't "surprise". You place what feels like a simple trade, the underlying barely moves, and your option price still bleeds. Or the stock goes your way and the option doesn't pay the way you imagined. Or volatility collapses after the event you were waiting for and your P&L gets punched in the throat even though you were "right."

The Greeks exist to remove surprise. Not by predicting the future, but by telling you what you're exposed to *right now*—the way a dashboard tells you whether you're speeding, whether the engine is revving too hard, and whether you're about to run out of fuel. If you ignore the dashboard, you can still arrive. You just tend to arrive smoking.

There's a reason platforms plaster Greeks right next to bid/ask and implied volatility: they're not academic trophies. They're operating metrics. Each Greek is a sensitivity—a small, local "if this changes a little, your option changes about this much" estimate. Local is the key word. Greeks are most honest when the world moves slowly and least honest when the world does what markets love to do: gap, lurch, and panic.

Four Greeks do most of the real work: Delta, Gamma, Theta, and Vega. Learn them as exposures first, math second.

Delta: the steering wheel (and the option's undercover identity)

Delta answers the most basic question: *how much does my option behave like the underlying?*

A delta of 0.50 means, roughly, that if the underlying moves \$1, the option price tends to move about \$0.50 *in the same direction*, all else equal. Calls have positive delta; puts have negative delta. So a put with delta −0.40 tends to gain about \$0.40 when the underlying drops \$1 (again: roughly, locally, all else equal).

The first way delta saves you is by preventing a common self-deception: thinking you "own upside" when you really own a tiny sliver of it. If you buy a far out-of-the-money call with delta 0.10, you didn't buy "the stock but cheaper." You bought something that currently behaves like about 0.10 shares per option share. That's not an insult; it's the trade. You're paying for convexity and optionality, not immediate participation.

The second way delta saves you is by revealing accidental stock exposure inside "volatility trades." A position that feels like a clever options structure can still have a delta that's basically screaming, "you're just long the stock". Delta is the lie detector for strategy narratives.

Delta also has a second life as a rough probability proxy: for many vanilla options, the absolute value of delta is often loosely related to the chance of finishing in-the-money under certain modeling assumptions. It can be a useful intuition pump, but it's not a promise. Markets don't owe you a probability just because the screen prints a number.

A practical habit: never enter an options position without knowing your net delta. If you can't say, "I'm effectively long about X shares," you're trading with the steering wheel disconnected.

Gamma: the accelerator (and why options can turn you into a riskier trader than you feel)

Gamma is delta's rate of change. It answers: *if the underlying moves, how much will my delta change?*

This is where options stop being "leveraged stock" and start being their own animal. Delta tells you the direction exposure now. Gamma tells you how quickly that exposure can transform as the underlying moves.

Long options tend to have positive gamma. That means when the underlying rises, a long call's delta increases (you become more long). When the underlying falls, the call's delta decreases (you become less long). That's a beautiful property: your exposure increases when you're right and decreases when you're wrong. It's why long options are called convex.

Short options tend to have negative gamma. That means your exposure often gets worse as the market moves against you: you can become *more* short as the underlying rises (short call), or *more* long as it falls (short put). Negative gamma is why short option positions can feel fine for weeks and then suddenly feel like they're sliding on ice.

Gamma also explains a classic emotional whiplash: "I was only a little wrong and then I was very wrong." That's delta changing under your feet.

Where does gamma live? Mostly near the money, and it becomes more intense as expiration approaches. That's why near-expiration at-the-money options can feel like they have superpowers: small underlying moves can change deltas fast. It's also why they can chew you up if you sell them casually.

If delta is your position's identity, gamma is its *mood swings*.

Theta: the rent you pay to hold possibility

Theta is time decay. It answers: *how much value does my option tend to lose (or gain) as one day passes, all else equal?*

This is the Greek that turns beginners into conspiracy theorists. You buy a call, the stock doesn't move, and the option price drifts down day after day. It feels unfair until you accept the nature of what you purchased: a right that expires. Time isn't neutral in options. It's an ingredient.

Long options generally have negative theta: time passing hurts you. Short options generally have positive theta: time passing helps you. That's why selling options is so tempting. You can make money without being "right" about direction, as long as nothing dramatic happens and volatility doesn't explode.

Theta is also not linear. All else equal, the decay tends to accelerate as expiration approaches, especially for options near the money. Think of it like ice melting: a big block seems fine for a while, then suddenly it's a puddle. The closer you get to expiration, the more you're holding a perishable asset.

There's a trader's version of this that's worth memorizing: *theta is what you pay for gamma*. If you want the convexity that lets your delta improve as the market moves, you generally pay for it with time decay. If you want to collect time decay, you often sell convexity and take on negative gamma. That trade-off is the heartbeat of options.

If you're holding long premium and you feel the clock pressuring you, that's not weakness. That's the contract working as designed.

Vega: the price of fear (and the reason you can be right and still lose)

Vega measures sensitivity to implied volatility (IV). It answers: *if implied volatility changes by 1 percentage point, about how much does my option price change?*

Implied volatility is the market's pricing of uncertainty. When IV rises, options tend to get more expensive because the market is assigning a higher likelihood to big moves before expiration. When IV falls, options tend to get cheaper because the market is assigning a lower likelihood to drama.

Long options generally have positive vega: rising IV helps you. Short options generally have negative vega: rising IV hurts you.

Vega is why earnings trades have such a strange vibe. Before earnings, IV often rises because everyone knows a move is coming. After earnings, IV often collapses because the uncertainty has been resolved. So you can buy a straddle, get the move you wanted, and still lose money if the volatility collapse is larger than the realized move you captured. That experience feels like betrayal until you recognize what you actually bought: not just movement, but *movement relative to what was priced in*.

Vega is also why options can behave differently across expirations. Longer-dated options usually have more vega than short-dated options, because there's more time for uncertainty to matter. That's one reason spreads like calendars exist: they're often bets on how IV is priced across time.

A useful way to keep vega from becoming mystical: treat it like a "headline sensitivity". If the market's fear gauge changes, how much does your position care?

Greeks are not independent: they move as a pack

One of the fastest ways to misuse Greeks is to treat each one like a separate dial you can adjust without consequences. In reality, they're entangled.

Near-the-money options often have:

- meaningful delta (you participate),

- high gamma (your delta changes quickly),
- heavy theta (the clock matters),
- substantial vega (IV changes matter).

Far out-of-the-money options often have:

- low delta (small participation),
- lower gamma until you get close to the strike,
- theta that can still be painful relative to premium,
- vega exposure that can dominate the price on quiet days.

Time to expiration shifts the balance. Volatility levels shift the balance. The underlying moving shifts the balance. Greeks are a snapshot of a living animal, not a tattoo.

So treat Greeks the way a pilot treats instruments: essential, constantly monitored, and always interpreted in context.

Net Greeks: stop thinking ``one contract," start thinking ``one position"

A single option's Greeks are interesting trivia. Your *position's* Greeks are risk management.

If you have multiple legs—a spread, a straddle, a hedge—your net delta, gamma, theta, and vega are what matter. You can have a position that is:

- delta-neutral (not directionally biased),
- gamma-positive (benefits from movement),
- theta-negative (pays rent daily),

- vega-positive (benefits if IV rises).

That's not a fantasy; it's a common long-vol structure.

Or the mirror image:

- delta-neutral,
- gamma-negative,
- theta-positive,
- vega-negative.

That's a classic short-volatility posture: collect decay, fear the sudden move.

Naming your net Greeks is a way of naming your enemies. If your position is short gamma, you should expect stress during fast markets. If your position is long theta-negative, you should expect boredom to be costly. If your position is short vega, you should expect implied volatility spikes to hurt even if the underlying hasn't moved much yet.

You don't want to be surprised by the identity of what can kill you.

A story markets tell over and over: the ``quiet bleed" and the ``violent snap"

Most option P&L stories fall into two genres.

Genre one: the quiet bleed. You buy options because you want convexity. The market refuses to move. Theta does its steady work. You start bargaining with the universe: "Just a little pop." It never comes. You learn that long options require not just being right, but being right *fast enough*. That's theta.

Genre two: the violent snap. You sell options because you want income. The market stays calm. Your P&L looks like competence. Then a gap happens. Delta jumps against you because gamma is negative. IV spikes and vega punches you too. The spread widens right when you want out. You learn that short options can be "small wins" until the day they are "large losses." That's gamma plus vega plus microstructure.

The Greeks are how you see these genres coming before the plot twist.

Using Greeks without getting hypnotized

Greeks are estimates, often computed from a model with assumptions (volatility surface shape, interest rates, dividends, and more). Different platforms can show slightly different values. That doesn't make them useless. It makes them *instruments*, not scripture.

Here's the practical way to use them:

- Delta: sizing and directional exposure. "How many shares am I effectively?"
- Gamma: path risk. "Will my delta stay tame or whip around if the underlying moves?"
- Theta: time budget. "How much am I paying/earning per day to hold this exposure?"
- Vega: volatility regime risk. "If fear gets repriced, do I benefit or get hurt?"

If you can answer those four questions, you're trading like someone who expects markets to be alive, not like someone hoping the mid price behaves.

Actionable takeaway

Before you enter any options trade (single-leg or spread), write a one-line "Greek label" for the *position*:

> *Net delta: ___ (shares). Net gamma: ___ (small/medium/large). Net theta: ___ (paying/collecting per day). Net vega: ___ (benefits/hurts if IV rises).*

Then add one operational rule based on the label:

- If you're short gamma, pre-commit to an adjustment/exit level before the market moves fast.
- If you're long theta-negative, set a time stop (a date) for re-evaluating, so you don't "marry" a decaying asset out of stubbornness.
- If you're short vega, decide what you'll do if IV jumps even without a big price move.

That's what Greeks are for: not to impress anyone, but to keep your trade from becoming a surprise story you tell later with the tone of someone who touched a hot stove "just to see."

2.6 Respecting the Leverage: Margin and Assignment

The most dangerous thing about options isn't the Greeks. It's the false sense of adulthood you get from a position with a neat payoff diagram.

A payoff diagram is honest *at expiration*. Your broker is honest *every minute before then*. That's the gap where blowups live. You can have a trade that is "defined risk" in theory and still get forcibly liquidated in practice. You can have a trade with a small maximum profit and still wake up to a surprise stock position because someone exercised against you overnight. You can be "right eventually" and still be wrong on Tuesday at 2:17 p.m., when your margin requirement spikes and your platform starts selling things you didn't mean to sell.

Margin and assignment are the unglamorous mechanics that turn options from a clever idea into a real-world instrument. This is where the market stops being a math problem and becomes plumbing, policy, and human behavior.

Margin: the broker's way of asking, ``Can you actually keep your promises?"

When you buy an option, you usually pay the premium up front. Your maximum loss is generally that premium. From the broker's perspective, the risk is simple: you already paid.

When you *sell* an option, you receive premium up front, but you also make an open-ended promise. If you're short a call, you might have to deliver shares at the strike price even if the stock rips higher. If you're short a put, you might have to buy shares at the strike price even if the stock collapses. The premium you collected is not collateral; it's a down payment on a liability.

Margin is the broker's way of ensuring you can make good on that promise *before* the bill arrives. It is not a moral judgment. It is not a suggestion. It is the broker doing risk management with *your* money on the line and *their* legal obligations in the background.

And here's the part most traders learn the hard way: margin is not static. It moves as the underlying moves, as volatility changes, and as your position changes. The moment your position becomes scarier, your required collateral can jump. That's when you discover the difference between "I think I can handle this trade" and "my broker agrees."

Reg-T vs portfolio margin: two philosophies of ``how bad could this get?"

Most retail accounts live under Regulation T (Reg-T) margin, a rule-based framework. Think of it as a standardized seatbelt: it doesn't know you personally, but it tries to keep you from going through the windshield. For options, Reg-T margin requirements are typically strategy-based and conservative, especially for naked short options.

Portfolio Margin is a different philosophy: it tends to estimate risk using scenario-based models that look at how your whole portfolio might behave across market moves. In the best case, it rewards diversification and hedging by reducing margin on offsetting positions. In the worst case, it gives you more rope precisely when your positions look "hedged" on paper but become correlated in a crisis. Portfolio margin can be a powerful tool, but it is also a sharper tool. Sharp tools cut both ways.

Two practical realities matter more than the label on your account:

- Brokers can impose requirements that are *stricter* than regulatory minimums, especially during volatile periods.
- Requirements can change quickly when markets are stressed, because the broker is trying to survive the same storm you are.

Margin is not just about you. It's about the broker's risk, the clearing system's risk, and the reality that markets sometimes gap in ways models don't politely anticipate.

The simplest way to think about margin: it's a volatility tax on short promises

If you want an intuition that's hard to forget: short options are promises that get more expensive to underwrite when the world looks unstable.

When implied volatility rises, the market is signaling higher expected movement. Your short option obligation becomes more dangerous, because the chance of a deep in-the-money outcome increases, and the mark-to-market swings get larger. Brokers react by asking for more collateral. Even if your final payoff is "defined" because you're in a spread, your *path risk* can still expand, and margin is about path risk as much as expiration risk.

That's why traders who sell options for income often feel a double hit in a selloff:

- The position loses money (delta/gamma moves against them).
- The margin requirement rises (volatility and risk increase), reducing free buying power.

This is how a portfolio can get squeezed: you're losing money and simultaneously losing flexibility.

Buying power is your oxygen tank

A lot of option blowups are not "I was wrong." They are "I ran out of oxygen."

Buying power (or excess liquidity) is the cushion between you and forced liquidation. It's what lets you hold through noise, adjust rationally, or roll a position without panic. When that cushion disappears, your strategy stops being about markets and starts being about survival.

This is why "high probability" short option strategies can be deceptively dangerous. They produce many small wins, which encourages size creep. Then one adverse move arrives, implied volatility spikes, and the broker demands more collateral. If you sized the trade assuming calm conditions, the margin expansion in stress can be the event that kills you, regardless of what happens by expiration.

If you want one rule that sounds boring and saves accounts: *never run your account close to maximum margin.* The market doesn't announce when it's about to become expensive.

Assignment: the market enforcing your obligations

Assignment is what happens when you are short an option and the option holder exercises. You don't get to vote. If you sold the contract, you agreed to the obligation. When exercise occurs, the clearing system assigns that exercise to someone short, and your broker allocates that assignment to an account according to its method.

Two features make assignment psychologically nasty:

- *Timing is uncertain.* For American-style options (common in U.S. equity options), exercise can occur on any business day before expiration.
- *It happens when you're not looking.* Many traders experience assignment as an overnight surprise: you go to sleep short a put and wake up long 100 shares per contract, or short a call and wake up short stock.

Assignment is not rare because markets are cruel. It's rare-ish because exercising early is often *not optimal* for the holder. But "often" is not "always," and your risk management should be built around "always."

The assignment you expect (expiration) vs the assignment that ambushes (early exercise)

At expiration, assignment feels normal. You know the deadline. You know which options are in-the-money. You can plan.

Early assignment feels like getting tackled from behind because it can occur when an option has little time value left and exercise makes economic sense for the holder.

The most common situations where assignment risk rises:

- Your short option is deep in-the-money and close to expiration.
- A short call is in-the-money ahead of an ex-dividend date, and the remaining time value is small relative to the dividend.

That second bullet is the classic trap. If you're short an in-the-money call on a dividend-paying stock, the call holder may exercise early to own the shares and collect the dividend. If they exercise, you get assigned and you may wake up short stock. If you don't have the capital, margin, or borrow availability to carry that short stock position, your broker may act quickly—and not gently.

Even worse, the assignment can create a position you never intended to hold *over a dividend date*, which can create additional cashflow effects.

The key emotional point: you can do everything "right" in your strategy logic and still get operationally punched if you ignore dividend calendars and time value.

Exercise, assignment, and ``pin risk": the weirdness near the close

Near expiration, especially on big option days, stocks can hover near strikes with large open interest. Traders talk about "pinning," but the real danger for you is more personal: *uncertainty about whether you'll be assigned.*

If your short option finishes very close to the strike at the close, you can end up in a twilight zone:

- You might think you're safe because it looks out-of-the-money.
- After-hours moves can change the economic incentive for the holder.
- Exercise decisions can be made based on final settlement rules and deadlines that don't match the vibe of your P&L chart.

This is called pin risk: you don't know your final position when the market closes, and you can end up unintentionally long or short stock Monday morning. For small accounts, that can be more dangerous than the option loss itself because the stock position can be large relative to capital.

If you trade near-expiration short options around key strikes, you are volunteering for this uncertainty. Make sure you're being paid for it.

Defined-risk spreads can still surprise you

A vertical credit spread looks like a seatbelt: max loss is capped. So why do traders still get burned?

Because the broker is managing risk in real time, and your spread is not immune to:

- widening bid/ask spreads (making exits expensive),
- assignment on the short leg while the long leg remains unexercised,

- temporary margin expansion because the broker is looking at the stock position created by assignment,
- and the simple fact that "max loss" can still be large relative to your account if you sized it like a lottery.

The classic headache is a short leg assigned early. Now you have stock exposure. Your long leg provides protection in theory, but it may not automatically exercise, and you may need to take action to flatten the position. Meanwhile, your margin requirement can change dramatically because you're holding stock (or short stock), not just a tidy spread.

This isn't an argument against spreads. It's an argument against treating defined-risk as "set and forget." The operational layer still matters.

The hidden villain: liquidity during stress

Margin calls and forced liquidations are not polite events that happen at fair prices.

They often happen when:

- volatility is elevated,
- spreads are wide,
- everyone is hitting bids,
- and your "mid price" is a fantasy.

This is why the earlier obsession with bid/ask spreads isn't pedantry. In stress, the spread becomes a second loss. You can be down on the position and down again on execution.

It's also why traders who survive tend to obsess over boring details like position size and liquid underlyings. Liquidity is a form of insurance you don't appreciate until you need it.

How to reduce assignment risk without becoming paranoid

You can't eliminate assignment risk if you're short American-style options. You can, however, stop being surprised by it.

Practical habits that help:

- If you're short in-the-money calls on dividend-paying stocks, know the ex-dividend date and check whether the remaining extrinsic value is small.
- Avoid letting short options sit deep in-the-money near expiration unless you *want* the stock position.
- Keep enough buying power to survive a temporary stock position created by assignment.
- Prefer liquid options where you can close or roll without paying a ransom spread.

None of this requires complex math. It requires respect for the fact that options are enforceable contracts, not just chart lines.

The ``math works until the broker liquidates you" checklist

If you trade options long enough, you'll meet someone who had a great thesis and a broken account. It's usually not because the payoff diagram lied. It's because the operational reality was ignored.

Before you sell premium, ask:

- *If volatility spikes, can I fund the margin increase without selling my best positions?*
- *If I'm assigned overnight, will the resulting stock position be survivable?*
- *If spreads widen, can I still exit without crossing a canyon?*

These questions don't make you timid. They make you durable.

Actionable takeaway

Adopt one non-negotiable operating rule for any strategy that involves short options:

> *Always size so that an overnight assignment or a volatility-driven margin increase is annoying, not catastrophic.*

Make it concrete. For every short option position (including credit spreads), do a quick "overnight stress test" in your notes:

- Assume you get assigned on the short leg.
- Convert that into the resulting stock position (e.g., 100 shares per contract).
- Ask whether your account could hold that position for a day without a margin call if the stock gaps another 5% against you.

If the answer is "no," the trade is too big or the underlying is the wrong arena. Options reward creativity, but they only reward it if you can stay in the game long enough for the creativity to matter.

Chapter 3

Cracking the Code: Realized vs. Implied Volatility

Imagine you run a car insurance company. Your profitability depends on two things: how many accidents actually happen (Realized Volatility) and how much you charged for the policies based on your fear of those accidents (Implied Volatility). If you charge for a hurricane but only get a drizzle, you keep the difference. This chapter isn't about complex math; it's about understanding that spread. We will decode the difference between the market's actual movement and the price tag the market puts on its own anxiety.

3.1 Measuring the Past: Realized Volatility

Two stocks can hand you the exact same one-year return and still feel like they came from different planets.

One glides upward with the calm confidence of an escalator. The other climbs the same distance by slipping on banana peels every few weeks—down 3%, up 4%, down 5%, up 6%—until, somehow, it ends at the same destination. If you only look at the start and end point, you'd call them identical. If you had to *live through* the ride, you'd swear one of them was trying to shake you out of the car.

That gap between *destination* and *ride* is what realized volatility is trying to capture.

Realized volatility (often shortened to *realized vol* or *RV*) is the market's speedometer *after the fact*. Not a prediction. Not a vibe. Not a "fear gauge." Just a measurement of how violently price actually moved over some window of time. The key word is *realized*: it already happened. You can argue about what it *means*. You can argue about what window to use. But you can't argue with the tape.

The useful mental model is simple: returns are the footsteps, and realized volatility tells you how erratic those footsteps were. Big, jumpy strides? High realized vol. Small, steady steps? Low realized vol. The reason traders obsess over this is also simple: once you measure the market's speed, you can size risk like an adult.

Volatility is path, not outcome

Most investors secretly think in outcomes: "I'm up 12%", "I'm down 8%", "I broke even." Your P&L statement trains you that way. Volatility ignores the score and watches the whole game film.

That matters because path changes behavior. A portfolio that dips 25% mid-year and finishes flat is not the same experience as a portfolio that drifts gently and finishes flat. One creates margin calls, panic selling, risk committee meetings, and the kind of emotional scarring that makes people miss the next rally. The other creates boredom—and boredom, in markets, is underrated alpha.

Realized volatility is how you put a number on that lived experience. It's the bridge between "This thing feels risky" and "This thing *is* risky, in measurable units."

The trader's definition: ``How wide is the daily wobble?"

If you want a definition that's more practical than academic, try this one:

Realized volatility is the typical size of daily percentage moves, expressed as an annualized number.

That sentence hides two important choices:

- *What counts as a "move"?* (Close-to-close? Intraday? Something else?)
- *What does "typical" mean?* (Standard deviation is the usual answer, but the intuition comes first.)

You don't need to start with a formula. Start with the idea of a daily wobble. If an index usually wobbles about 0.5% per day, it's a sleepy market. If it wobbles 2% per day, it's a market where stops get hit for sport.

Traders like to annualize that daily wobble because it makes different assets comparable. A 2% daily wobble is hard to compare across products and time horizons; an "annualized vol" number lets you talk in one standardized language. It's imperfect, but it's a lingua franca.

Close-to-close realized vol: the classic, blunt instrument

The simplest realized volatility measure uses daily *close-to-close* returns.

Take a series of closing prices, convert them into daily returns, measure how spread out those returns are, then scale it up to an annualized number. That's the whole game.

Why close-to-close?

- The data is clean and ubiquitous.
- It lines up with how many funds mark their books (end-of-day).
- It's easy to compute and easy to communicate.

Why is it blunt?

- It ignores the drama that happens during the day.
- It can miss the "round trip" days where price swings wildly and then closes near unchanged.
- It mashes together two different beasts: overnight gaps and regular-session movement.

Close-to-close vol is like judging a movie by the first and last frame. Sometimes that's enough. Sometimes you miss the car chase.

A quick detour: returns that behave better than prices

Volatility is almost always computed on *returns* (percentage changes), not raw price changes. That's because a $5 move in a $50 stock is a different animal than a $5 move in a $500 stock. Returns normalize the move.

Many practitioners use log returns (the change in the natural log of price) because they add nicely over time and behave a bit better statistically. You don't have to become a log-return zealot to be effective; the main point is consistency. If you're comparing realized vol across assets or across time, measure returns in the same way.

Standard deviation: the ``typical move" with sharp edges

When people say "volatility," they usually mean standard deviation of returns. Standard deviation is a way of quantifying typical dispersion: if returns are tightly clustered, standard deviation is low; if they're spread out, it's high.

Here's the important emotional truth standard deviation captures: *surprise costs money.* The wider the distribution of daily returns, the more often you get punched in the face by a move you weren't sized for.

But standard deviation has sharp edges. It treats upside and downside moves symmetrically, even though humans don't. It also assumes a kind of bell-curve world where extreme moves are rare—and markets, as you've probably noticed, love extreme moves. Still, it remains the workhorse because it's simple, comparable, and deeply embedded in option pricing and risk systems.

As a volatility trader, you don't worship standard deviation. You use it the way a pilot uses an altimeter: you know it's not the entire sky, but you'd be insane to ignore it.

The square-root-of-time trick (and why 252 shows up everywhere)

At some point you'll see a daily volatility number multiplied by $\sqrt{252}$ to get an annualized number. This looks like finance numerology until you understand the assumption hiding underneath:

If daily returns are independent and have a stable variance, variances add over time.

Variance is volatility squared. If variance adds linearly with time, volatility adds with the *square root* of time. That's why you scale by $\sqrt{N}$, not by N.

Why 252? Because there are about 252 trading days in a year.

So the practical workflow looks like this:

1. Compute daily returns.
2. Compute the standard deviation of daily returns (daily vol).
3. Annualize: $\sigma_{\text{annual}} \approx \sigma_{\text{daily}} \times \sqrt{252}$.

This scaling is incredibly useful—and also easy to abuse. Markets are not perfectly independent day to day. Volatility clusters. Crashes create serial correlation in turbulence. The square-root-of-time rule is a *translation tool,* not a law of physics.

Still, as translations go, it's one of the best we have. It lets you take something you can feel (daily chop) and express it in the common unit used by options markets (annualized volatility).

Lookback windows: the market's memory is your choice

Realized volatility always comes with a lookback window: 10-day, 20-day, 63-day, 100-day, 1-year. Each window answers a different question.

A 20-day realized vol is essentially asking: "How wild has this been *lately*?" It reacts quickly. It will spike after a few chaotic sessions and calm down quickly after the storm passes.

A 100-day realized vol is asking: "What's the *typical regime* recently?" It moves slower. It won't overreact to a brief tantrum, but it can also stay elevated long after the market has gone back to sleep.

Here's the trap: people treat realized vol as a single truth, when it's really a family of truths. Change the window and you change the story. That's not a bug. It's the point.

A useful way to think about it:

- Short windows are better for *tactical* decisions (position sizing, near-term hedging).
- Longer windows are better for *baseline* context (is current vol high relative to the last year?).

If you're trading volatility, you're not just measuring movement. You're measuring whether the market has entered a new mood.

Close-to-close misses the ``intraday zig-zag"

Now for the part that surprises people: you can have a day with a tiny close-to-close return and enormous realized volatility *inside* the day.

Imagine the S&P opens at 100, rips to 103 by lunch, dumps to 98 by mid-afternoon, then claws back to close at 100. Close-to-close return: 0%. Emotional experience: a stress test. Risk experience: stops triggered, hedges whipsawed, liquidity pockets exposed.

Close-to-close realized vol undercounts that day because it only sees the start and the finish. For some purposes (like end-of-day marking), that's fine. For others (like understanding how violently the market traded), it's like measuring an earthquake by where the furniture ends up rather than how hard the building shook.

This is where intraday measures enter the picture.

Intraday realized volatility: measuring the shaking, not just the closing print

If you have intraday prices, you can compute returns over shorter intervals: 5-minute, 15-minute, 30-minute. Then you can aggregate those small moves into a daily realized variance measure.

The intuition is elegant: if the market's path matters, sample the path more frequently.

This tends to produce a more informative measure of what actually happened during the day, especially on "round trip" sessions. It can also help separate two different sources of turbulence:

- *Overnight risk:* gaps from the close to next open, driven by news, earnings, geopolitics.
- *Intraday risk:* the tug-of-war while the market is open, driven by flows, positioning, and liquidity.

That split is practical. Overnight gaps can't be stopped out intraday. Intraday swings can. If you run a strategy that trades during market hours, you care deeply about intraday realized vol. If you hold risk overnight, you care about both.

There's also a subtle point: intraday realized vol can "see" volatility clustering faster. When the market transitions from calm to chaotic, the first signs often show up as wider intraday ranges before the closes start printing big moves.

The high-frequency caution: more data isn't always cleaner

More frequent sampling sounds like free truth, but there are landmines.

At very high frequency (tick data, second-by-second), price moves are polluted by microstructure noise: bid-ask bounce, discrete pricing, temporary illiquidity. That noise can inflate volatility estimates if you're not careful.

That's why many practitioners use moderate intervals (like 5-minute returns) rather than the tiniest granularity. You want enough resolution to capture the path, but not so much that you start measuring the market's plumbing instead of its motion.

If you're reading this as a trader, not a quant, the takeaway is simple: intraday realized vol is powerful, but treat it like a microscope. Zoom in too far and you stop seeing the organism.

Realized variance vs realized volatility: why the square matters

You'll hear two closely related terms:

- *Realized volatility:* the square root of realized variance.
- *Realized variance:* roughly the sum of squared returns over the window.

Variance is "volatility squared," which feels abstract until you realize why options people like it: option payoffs are convex, and variance is a natural currency for convex risk. Many volatility products, swaps, and academic results are expressed in variance terms because it aggregates neatly and links cleanly to certain replication arguments.

For most discretionary traders, volatility is the more intuitive unit because it speaks the language of "percent moves." But when you start comparing realized movement to what options were pricing, thinking in variance terms can keep you honest. Squaring returns makes big moves count a *lot* more, which matches how drawdowns and crash risk dominate P&L.

Why realized vol is a risk tool before it's a trading signal

A common beginner mistake is to treat realized volatility like a forecast: "Realized vol was 10, so next month will be 10." Markets love punishing that sentence.

Realized vol is best used as:

- a *speed check* (how fast has this market been moving?),
- a *regime clue* (has the character changed recently?),
- and a *position-sizing input* (how much risk can I carry without getting randomly obliterated?).

Even if you never trade a single option, realized vol belongs in your toolkit because it turns risk from a feeling into a number.

Consider two position-sizing philosophies:

- *Fixed shares:* buy 100 shares no matter what.
- *Vol-adjusted:* buy fewer shares when realized vol is high, more when it's low.

The first approach accidentally takes more risk precisely when the market is most dangerous. The second approach is an attempt at sanity: if the road is icy, drive slower.

Volatility targeting and risk parity strategies are basically industrial-scale versions of this common sense. They don't eliminate risk, but they reduce the chance you wake up in a new regime with the same position size you used in a calm one.

The ``same vol, different risk" problem: distributions matter

Here's a twist that keeps realized vol from being the whole story: two return series can have similar standard deviation but very different *shape*.

One can have frequent medium moves in both directions. The other can have lots of tiny days and a few monstrous gaps. Both can produce the same realized vol number. Only one of them blows up short-vol traders.

This is where concepts like skewness (are the big moves mostly down?) and kurtosis (how fat are the tails?) start to matter. You don't need those words to grasp the reality: *how* volatility arrives is part of risk.

Realized vol is still the right starting point because it's measurable and comparable. Just don't confuse "volatility" with "danger." Volatility is the *amplitude*. Danger is amplitude plus the probability the next move is a trap door.

A practical way to compute realized vol (without turning it into homework)

If you want a clean routine you can actually use, here's a practical recipe:

1. Pick a lookback window (start with 20 trading days).
2. Compute daily returns from close to close.
3. Compute the standard deviation of those returns.
4. Annualize by multiplying by $\sqrt{252}$.

Then do something even more useful: compute a second realized vol with a longer window (say 100 days) and compare them.

When the 20-day vol is far above the 100-day vol, the market has recently gotten louder. When the 20-day is far below, the market has recently calmed down. That spread is a quick-and-dirty regime detector you can glance at before you place a trade.

If you have intraday data, you can add an intraday realized measure later, but the close-to-close version is the right default for building intuition. Crawl before you sprint.

Actionable takeaway

Before putting on any trade that can lose real money quickly (options, leveraged ETFs, concentrated single-stock bets), compute two realized volatility numbers on the underlying: a 20-day and a 100-day close-to-close annualized vol. If the short window is meaningfully higher than the long window, assume you're in a louder regime than your recent memory suggests and cut size *before* the market forces you to. If the short window is meaningfully lower, don't automatically size up—but recognize the hidden risk: calm markets tempt you into complacency, and complacency is where blowups are born.

3.2 Pricing the Future: Implied Volatility

If you've ever watched an options chain light up during a market wobble, you've seen a very human reflex in numeric form: people don't buy options because they love math. They buy them because uncertainty feels awful, and paying a premium to make that feeling go away can seem like a bargain.

That premium has a name traders use like it's a physical substance: implied volatility.

Implied volatility (IV) is *not* a forecast handed down from Mount Quant. It's the volatility number that makes an option pricing model "fit" the price the market is actually trading at. In other words, IV is the market's receipt. It's what the crowd collectively paid for uncertainty, translated into the familiar unit of annualized volatility.

That framing changes your posture immediately. If IV is a receipt, then it's not asking "What will happen?" It's telling you "What did everyone just agree to pay for protection or leverage *right now*?"

The umbrella problem: price responds to fear faster than weather

Imagine you run into a convenience store because the sky looks like it's about to open up. Umbrellas were $10 yesterday. Today, they're $35. Did the umbrella seller discover a meteorological truth about *exactly* how many inches of rain will fall?

No. The price jumped because demand surged in the moment and supply didn't.

Options behave the same way. When markets get twitchy, the demand for downside insurance (puts) and convexity (that sweet, curved payoff) often spikes. The supply of people willing to sell that insurance doesn't magically expand at the same speed. So the price of options rises. When you run that higher option price through a model, the volatility input that makes the model match the market price also rises. That backed-out number is implied volatility.

IV is a *clearing price for uncertainty*. It's the equilibrium between buyers who want protection and sellers who want to be paid for bearing risk.

And like umbrellas, it can get expensive even if the storm ends up being brief.

Implied volatility is a plug variable, not a prophecy

Option pricing models take several inputs: the current price of the stock (or index), the strike, time to expiration, interest rates, and an assumption about volatility. In textbooks, volatility is treated like an input you *know*. In markets, volatility is the input you *don't* know—so you back it out from the price.

That's the inversion that makes IV so easy to misunderstand. People see a number (say, 18%) and assume it's a forecast. But it's more like the implied yield on a bond: it's a rate that's embedded in a traded price. It tells you what the market is pricing, not what reality will deliver.

Here's the most honest way to say it:

Implied volatility is the volatility that would make the option's theoretical price equal its market price, given the model and other inputs.

That "given the model" clause matters. Change the model's assumptions, and the implied number you back out can shift. Yet, despite its imperfections, IV is still enormously useful because the market coordinates around it. It's the common language traders use to compare option richness or cheapness across strikes and expirations.

Place this right after the "plug variable" idea:

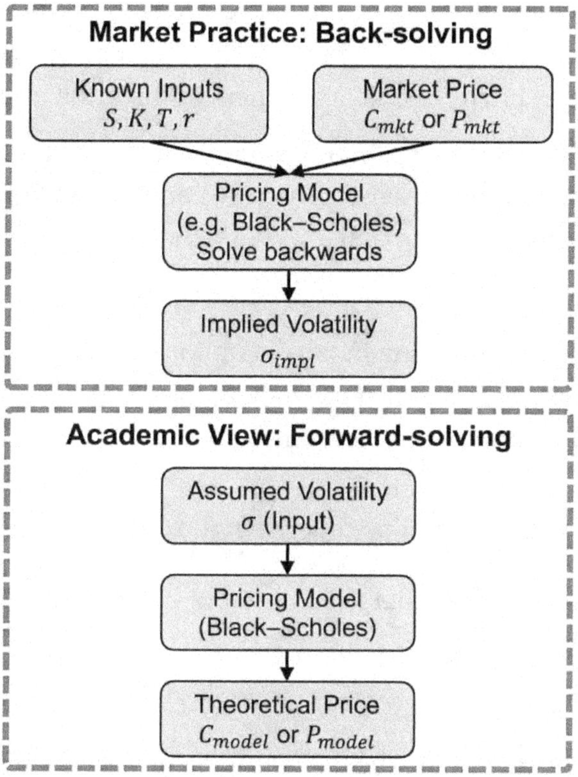

Why traders quote vol instead of dollars

If IV is backed out from price, why not just talk about the price? Because option prices don't travel well across strikes, maturities, and underlyings. A $2 premium can be cheap or expensive depending on the stock price, the time to expiry, the strike, and how jumpy the underlying usually is.

IV normalizes all of that into one comparable number. It's like quoting airline tickets in "dollars per mile" rather than "dollars per seat." Not perfect, but much more informative.

Two practical reasons IV becomes the native tongue:

- *Comparability:* An at-the-money (ATM) option on a $50 stock and an ATM option on a $500 stock live in different dollar universes. Vol puts them in the same unit.
- *Risk management:* Many option books are managed in terms of vega (sensitivity to IV) and variance exposure, not in "premium dollars." Vol is the control knob traders actually turn.

Once you start thinking in vol, an options chain stops looking like a list of prices and starts looking like a market of beliefs, fear, and inventory management.

So what *is* the market pricing when it prices IV?

Here's the sentence that keeps you out of trouble:

Implied volatility is a price of risk under the market's rules, not a clean estimate of future realized volatility.

Those "rules" include risk aversion, leverage constraints, hedging flows, and the fact that losses hurt more than gains feel good. In options, that emotional asymmetry shows up as structural demand for protection, especially in equity indexes.

So even if traders were perfect statisticians (they're not), IV would still embed a premium for being short convexity—for being the person who sells umbrellas during a storm and promises to deliver them even if the sky falls.

That's why IV often trades above realized volatility over time. Not because the market is dumb, but because *insurance usually costs more than the average claim.* It has to, or nobody would write the policies.

IV is forward-looking, but not directional

Another misconception: "High IV means the market expects a crash." Sometimes a crash is part of the story, but IV itself isn't a directional forecast. It's mostly about the magnitude of potential movement.

An options market can price high IV in both directions: a huge rally is still a huge move, and gamma doesn't care about your politics. That said, equity markets have a persistent asymmetry: downside tail risk is more violently priced than upside tail risk. That shows up more in *skew* (how IV changes across strikes) than in the single IV number at the money.

Think of IV as a statement like: "Seatbelts are expensive today." That tells you people are nervous about turbulence. It doesn't tell you whether the plane will drop 500 feet or climb 500 feet.

The dealer's problem: inventories, hedges, and the cost of being short panic

If you want to understand why IV moves the way it does, don't picture a serene market making a probabilistic forecast. Picture a room full of professionals managing risk in real time.

Dealers (market makers) often end up on the other side of customer flows. When customers rush to buy puts, dealers sell those puts, and now dealers are short convexity. As the market falls, those short-put positions get more sensitive. Dealers typically hedge by selling the underlying as it drops (because their delta changes). That hedging can add fuel to the move, especially in fast markets.

This creates a feedback loop:

- Customers buy protection.
- Dealers sell it and hedge dynamically.
- Hedging pressure can amplify moves.
- Amplified moves increase demand for protection.
- Prices rise, and IV rises with them.

You don't need to believe this loop explains everything. You just need to see why "IV is the market's forecast" is too innocent. IV is also the cost of warehousing risk when everyone wants the same thing at the same time.

Why IV spikes fast and bleeds slow

Watch IV through any panic and you'll notice a pattern: it tends to gap higher like a heart rate monitor when the patient sits up, then drift lower as the fear metabolizes.

That behavior is not an accident. Fear is urgent; relief is gradual.

When uncertainty hits (a surprise CPI print, an earnings shock, a geopolitical headline), option buyers don't negotiate politely. They pay up because the alternative is being unhedged into a headline-driven gap. Option sellers, meanwhile, don't suddenly become brave. They demand a higher premium to take the other side. So IV spikes.

When the market calms down, the urgency fades. Sellers re-enter. Hedgers unwind. The bid for protection softens. IV "bleeds."

This is one reason short-vol strategies can look like a money machine for long stretches: collecting premium in quiet periods feels like picking up nickels that reset every day. The catch is that the steamroller doesn't send calendar invites.

IV as a range quote: turning a percent into a daily move

IV is annualized. Your portfolio is not. So the useful question becomes: "What does an IV of 20% *mean* for tomorrow?"

A quick-and-dirty translation is square-root-of-time logic: daily one-standard-deviation move is roughly

$$\text{Daily } 1\sigma \approx \frac{\text{IV}}{\sqrt{252}} \approx \frac{\text{IV}}{16}.$$

So 16% IV implies about a 1% daily move (one standard deviation). 32% IV implies about a 2% daily move. That's not a promise; it's a scale.

This translation is extremely practical when you're staring at an option chain and trying to decide whether you're paying a sane price for protection. If the market is pricing a 2% daily move and the last month has been drifting at 0.6% a day, you're not just buying insurance—you're buying it in a market where everyone else is grabbing umbrellas too.

A subtlety that matters: IV is "risk-neutral," not "most likely"

Even if IV were a perfect reflection of what the market expects, it would still be the expectation under *risk-neutral* pricing—a framework that bakes in risk preferences and the cost of hedging. That's why options can be "expensive" relative to realized outcomes and still be correctly priced in a market sense.

In plain English: the options market doesn't just price what might happen. It prices how painful it would be to be short protection if it happens, and how hard it is to hedge when things get ugly.

That pain factor is why the same realized volatility can coexist with very different implied volatility regimes. A sleepy market with fragile liquidity can trade higher IV than its realized movement would suggest, because the crowd knows that if the music stops, the exits are narrow.

The most practical way to use IV: treat it like a market-implied budget

Here's a clean way to use IV without turning it into mysticism:

IV is the market's budget for future movement, plus an insurance markup.

If you're an option buyer, you're paying that budget. If you're an option seller, you're collecting it and taking responsibility for the tail risk.

So you ask two grounded questions:

- *What is the market charging for movement?* (the IV level)
- *What kind of movement is the market most afraid of?* (skew across strikes, especially puts)

If IV is cheap, it doesn't mean nothing bad will happen. It means the market is offering you insurance at a discount relative to its recent mood. If IV is expensive, it doesn't mean disaster is imminent. It means you're shopping during a storm.

Actionable takeaway

The next time you look at an option quote, force yourself to say the sentence out loud (or at least in your head): "This implied volatility is a price, not a prophecy." Then convert the IV into a rough daily 1σ range using IV/16, and compare that to what you've actually been seeing in day-to-day moves. If the market is pricing a daily range that's wildly larger than recent reality, assume you're paying a fear premium and demand a better reason to buy. If it's pricing a daily range that's oddly small, assume complacency is in the air and treat cheap options like discounted fire insurance—not because a fire is guaranteed, but because the price is telling you the neighborhood is feeling a little too safe.

3.3 Model Intuition: Thinking Like a Pricer

A dealer once described options trading to me with the kind of dry humor you only earn after a few market panics: "I don't predict. I *balance*." That's the right mental switch. If you approach options like a fortune teller, you'll stare at implied volatility as if it's whispering secrets. If you approach options like a pricer, you start asking a better question: *What combination of trades makes this risk tolerable to hold?*

Black–Scholes is famous because it gives you a clean price from a clean formula. It's also famous because markets refuse to behave cleanly. The useful part isn't memorizing the equation; it's understanding the logic underneath: the model is built around the idea that you can manufacture (or "replicate") an option using the underlying stock plus borrowing/lending. Once you believe replication is possible, the option's price stops being a matter of opinion and becomes a matter of *no-arbitrage*: if the option is mispriced, someone can lock in a riskless profit by trading the replicating ingredients.

That's the heartbeat of modern derivatives pricing. And it's also where the cracks begin.

The core trick: turn an option into a hedgeable process

Picture what an option is from a dealer's perspective. It's not a lottery ticket. It's a position whose risk changes as the underlying price moves. A simple call option becomes more "stock-like" as the stock rises and less "stock-like" as the stock falls. That changing sensitivity is why options feel magical to retail traders and feel like a risk management problem to professionals.

The first Greek you need for intuition is *delta*. Delta is the option's sensitivity to the underlying: roughly, how many dollars the option changes when the stock moves one dollar. A call might have a delta of 0.50, meaning it behaves like about half a share of stock. A put might have a delta of -0.50. Delta is not constant; it moves with price, time, and volatility.

Now the replication idea: if you own an option, you can hedge some of its risk by taking the opposite amount of stock equal to its delta. If your option has delta 0.50, you short 0.50 shares of stock (or buy if you're short the option) to neutralize the immediate directional exposure.

The slogan is: *option risk can be traded into stock risk.*

But you don't get to do it once and go to lunch. Because delta changes.

Dynamic hedging: the ``self-updating" recipe that makes pricing possible

Black–Scholes leans on dynamic hedging: as the stock price moves, you continuously adjust your stock position to keep your option hedge aligned. In the model's world, you can rebalance instantly, in any size, at a fair price, without friction. That's not how real markets work, but hold that thought.

If you can rebalance continuously, something remarkable happens: you can eliminate the option's randomness over tiny time steps. The option's value wiggles around, but your hedge wiggles in the opposite way. If the hedge is perfect, the combined position earns the risk-free rate, like a bond. And if a position is riskless, competition forces its return to be the risk-free rate. That pins down a unique fair price for the option.

This is why the model feels so authoritative: it's not pricing based on "expected returns." It's pricing based on "no free lunch if you can hedge."

That logic is deeply practical because dealers actually do something that rhymes with it. They hedge. They don't sit around hoping a call goes up; they manage a book by neutralizing exposures and charging spreads for the risk they can't neutralize perfectly.

The hidden assumption: markets move like mist, not like rocks

The original Black–Scholes framework assumes the stock price evolves smoothly in a particular statistical way: lots of tiny random moves, no sudden jumps. If you like metaphors, it treats markets like particles floating in a gas: jittery, continuous, and always tradable.

That assumption isn't "wrong" in the sense that prices do move a lot by small increments. The problem is that the most important moves—the ones that define risk—often aren't small. They're discontinuities: earnings gaps, flash crashes, overnight geopolitical jolts, a central bank headline that hits like a cymbal.

Those are jumps. And jumps are kryptonite for the clean replication argument.

Why? Because dynamic hedging relies on being able to adjust *after* the price moves, in tiny increments. A jump skips the increments. Your hedge is sized for the world one millisecond ago, and then the market teleports. You can't rebalance in the void between prices. The option's value changes discontinuously, and your hedge can't offset it. The "riskless" portfolio is suddenly very risky.

That's the structural flaw: the model's elegance depends on a world without trap doors. Markets, unfortunately, install trap doors as a feature.

Volatility in the model vs volatility in the wild

In the Black–Scholes world, volatility is a constant parameter. In the real world, volatility is a living thing that gets agitated, calms down, then gets agitated again. It clusters. It spikes. It mean reverts. It is famously not constant.

This matters because dynamic hedging performance depends on volatility. If volatility jumps higher, the option's delta changes faster; you have to trade more to stay hedged. More trading means higher transaction costs and more slippage. In crisis conditions, liquidity thins and spreads widen—exactly when the model asks you to rebalance most aggressively.

So even if prices were continuous, hedging would still be imperfect because you don't rebalance in a frictionless vacuum. You rebalance in a real market with bid-ask spreads, latency, market impact, and the occasional moment where nothing is quoted at the size you need.

A good way to say it: *the model assumes hedging is easy when hedging is hardest*.

Gamma: why dealers end up trading *against* the move

If delta is the option's first sensitivity, *gamma* is the second: how fast delta changes as the underlying moves. Gamma is what makes options convex. It's also what makes hedging behave like a strategy.

When you're *long gamma* (typically long options), your delta increases as the market goes up and decreases as the market goes down. If you're hedging to stay delta-neutral, that forces you to *sell into rallies* and *buy into selloffs*. That's naturally stabilizing: it's "buy low, sell high" as a mechanical consequence of the hedge.

When you're *short gamma* (typically short options), the opposite happens. Your hedge forces you to *buy into rallies* and *sell into selloffs*. That's pro-cyclical. It can amplify moves, and in fast markets it can feel like the hedging tail wagging the price dog.

This is where "thinking like a pricer" becomes more than philosophy. It explains real market texture. During a selloff, if the street is net short gamma, hedging flows can accelerate the move and push implied volatility higher. Not because anyone predicted the selloff, but because the microstructure of hedging reacts to it.

Black–Scholes doesn't "cause" this, but the hedging logic it popularized helps you see why volatility can behave like a feedback system rather than a static statistic.

Vega: the Greek that makes IV feel like a second underlying

Earlier, implied volatility was framed as a price—a clearing price for uncertainty. Vega is the bridge from that abstract idea to your P&L.

Vega is how much an option's price changes when implied volatility changes by one percentage point (all else equal). If a call has a vega of 0.10, then a $+1$ vol point move in IV increases the option's theoretical price by about \$0.10. A $+5$ point move adds about \$0.50.

This is why traders say things like "I'm long vol" or "I'm short vol." They're not talking about owning a volatility index; they're talking about exposure to IV repricing through vega.

The most important practical insight: you can lose money on an option even if the stock doesn't move, purely because IV dropped. And you can make money even if the stock goes nowhere, purely because IV rose. Options have two engines: underlying movement and volatility repricing. Many beginners only watch one dashboard.

Vega also explains why longer-dated options often feel "stickier" and more expensive in vol terms: they tend to have higher vega. They are more sensitive to changes in IV because there's more time for uncertainty to matter.

Time: why the model's clean clock creates messy incentives

Black–Scholes treats time as a smooth continuum, like sand in an hourglass. Real markets have cliffs: earnings dates, economic releases, Fed meetings, option expirations. Uncertainty doesn't decay smoothly into those events; it piles up and then gets released.

That's why IV often behaves like a coiled spring around catalysts. The model's constant volatility assumption can't naturally represent "volatility scheduled for Tuesday morning," but the market can. It prices the event risk into options expiring after the catalyst and often leaves shorter expirations relatively calmer until the event comes into view.

Thinking like a pricer means noticing that options are not just about "30 days of movement." They're about *which* 30 days. A month that includes earnings is not the same as a month that doesn't, even if the calendar count is identical.

Why the model remains useful even when it's wrong

Here's the paradox: Black–Scholes is structurally flawed, and yet it's indispensable.

It's indispensable because it gives you a consistent way to translate between prices and vol, and a consistent way to talk about risk. It gives you Greeks, and Greeks let you decompose an option position into interpretable exposures: delta (direction), gamma (convexity), vega (vol pricing), theta (time decay).

Even when the model's assumptions are violated, the *sensitivities* can still be good approximations locally. That's how most practitioners use it: not as scripture, but as a map. A map doesn't need to include every tree to keep you from getting lost.

The danger is not using an imperfect model. The danger is forgetting it's imperfect and sizing trades as if the world will politely honor your assumptions.

Where the model breaks in the real world (and how that shows up in IV)

The structural flaws aren't abstract; you can see them in option surfaces.

If the model were perfectly describing reality, implied volatility would be roughly the same across strikes for a given expiration. In practice, equity index options typically show *skew*: downside puts trade at higher implied vol than upside calls. That's the market screaming, in a single picture, that "big down moves are more feared (or more expensive to insure) than big up moves."

Skew is the market's admission that returns aren't symmetric and that jumps matter. It's also an admission that hedging is more painful in crashes than in rallies. Liquidity disappears, correlations spike, and the cost of carrying short convexity goes through the roof. So the implied volatility you back out is not just an expectation of movement; it's a composite of expected movement plus the price of being short the tail.

Even at-the-money IV carries some of that DNA. It's why IV can stay elevated after realized vol calms down: the market is still charging for the possibility that the trap door is there, even if it hasn't opened recently.

A pricer's checklist: what matters when you're not pretending the model is reality

Thinking like a pricer means you stop asking "Is IV right?" and start asking "What risks does this price compensate me for, and which ones does it *not*?"

Some of those risks are hedgeable most of the time (small directional moves). Others are only partially hedgeable (fast moves, liquidity gaps). Others are essentially unhedgeable (overnight jumps, sudden correlation shocks). The option price has to compensate whoever holds that unhedgeable stuff, because no formula can wish it away.

This is the psychological shift that separates casual option buyers from consistent volatility traders: you trade the *market for risk transfer*, not the market for predictions.

Actionable takeaway

Before you trade any option position, write down (literally, on paper or in a note) which Greek you're actually betting on: *delta* (direction), *gamma* (convexity), or *vega* (IV repricing). Then ask one brutally practical question: "If the underlying doesn't move, do I still have a way to make money?" If the answer is "no," you're more directionally exposed than you think. If the answer is "yes, because IV can change," then you're trading the market's price of uncertainty—and you should size the position as if the model's clean hedging assumptions can fail precisely when you need them most.

3.4 The Edge: The Volatility Risk Premium (VRP)

There's a strange job in markets that almost nobody admits they have: *selling peace of mind.*

If you own a big equity portfolio and you're doing well, your "problem" isn't that you'll miss the next 2% rally. Your problem is that a fast, ugly selloff can erase a year of good decisions in a week, right when you least want to make decisions. So you buy protection. Not because you love paying premiums, but because you hate the alternative: explaining a drawdown to your future self, your spouse, your boss, your investors, or your risk committee.

That protection is mostly delivered through options. And the price of that protection is usually not cheap. The persistent gap between what options *imply* about future volatility and what the market later *realizes* is the volatility risk premium—VRP for short.

It's one of the most important "edges" in volatility trading, not because it's secret, but because it's structural. The incentives that create it are baked into human behavior and institutional constraints. The edge is real. So is the risk for anyone who tries to harvest it carelessly.

The clean definition (and the less clean reality)

Start with the simple version:

VRP is the tendency for implied volatility to be higher than subsequent realized volatility.

In the equity index world, that often shows up as VIX (a 30-day, option-implied volatility measure) sitting above the market's later 30-day realized volatility.

The more precise version uses variance instead of volatility:

Variance risk premium is the difference between risk-neutral expected variance (implied by option prices) and expected (or later realized) variance in the real world.

Why variance? Because option replication arguments naturally live in variance space, and because big moves matter disproportionately when you square returns. But you don't need to talk like a derivatives desk to use the concept. The intuition survives translation:

Options usually price in more turbulence than the market ends up delivering, because the market charges an insurance markup.

That "usually" is doing a lot of work. The premium exists on average, over time. In the wrong week, realized volatility can leap above implied volatility like a monster jumping out of a closet. That's not a bug in VRP. That's the bill coming due.

Why the premium exists: crash aversion is a business model

If you want to understand VRP, stop thinking about options as clever bets and start thinking about them as insurance contracts.

Insurance has a weird economic feature: buyers are often willing to overpay relative to the average claim, because the claim arrives when they are least able to pay it themselves. That's not irrational; it's survival.

Equity crashes are like that. They're not just "a negative return." They're correlated with job insecurity, funding stress, margin pressure, and forced selling. The timing is hostile.

So many investors don't buy puts because the expected value is positive. They buy puts because they're trying to avoid ruin, embarrassment, or career risk. That creates persistent demand for protection.

On the other side, the natural sellers of options (dealers, market makers, volatility funds, carry traders) are not philanthropists. They'll sell insurance, but only at a price that compensates them for:

- the possibility of a jump that can't be hedged cleanly,
- liquidity evaporating exactly when hedging is needed,

- and the unpleasant fact that short-option P&L tends to be "many small wins, occasional catastrophic loss."

That last point matters more than most people admit. Humans hate negatively skewed payoffs. A strategy that makes $1 most months and loses $20 occasionally is psychologically brutal and institutionally fragile. To run it, you need either a strong stomach, a strong mandate, or a strong fee structure. The market compensates that discomfort with a premium.

That premium is VRP.

The most important sentence in volatility trading

Here it is:

Implied volatility is a price for risk transfer, not a neutral forecast of realized volatility.

Once you internalize that, VRP becomes less mysterious. Of course implied tends to exceed realized. Of course the market overpays for protection. That's what insurance looks like in almost every domain where the downside is lumpy and painful.

What's special about options is that the "insurance premium" is quoted in volatility points, so it tempts people into thinking it's a scientific estimate. It's not. It's a negotiated price that reflects fear, constraints, and the difficulty of hedging tails.

Where VRP shows up in your P&L: short vol carry

VRP is the economic reason that short-volatility strategies exist at all.

If options were perfectly fairly priced in the everyday sense—if implied matched subsequent realized, with no markup—then systematically selling options would be like running an insurance company that charges exactly the expected claims. You might tread water before costs, and you'd sink after costs.

But with VRP, the seller is often paid a bit extra: implied exceeds what is later realized, so the option premium collected tends to be rich relative to the volatility the market actually delivers. That "extra" doesn't arrive as a neat check labeled *VRP*. It arrives as:

- theta (time decay) earned while nothing terrible happens,
- IV mean reversion after fear spikes,
- and the repeated experience of "I sold expensive uncertainty and the future wasn't as dramatic as the price implied."

It's also why the temptation is so strong. Selling options can feel like renting out shovels in a gold rush: consistent demand, steady cash flow, lots of satisfied customers. Until the day everyone needs shovels at once and the supply chain breaks.

VRP is not free money; it's payment for bearing ugly risk

If you've ever heard someone say "Options are overpriced," they're often pointing at VRP. But "overpriced" is a loaded word. Overpriced relative to what?

Relative to *subsequent realized volatility*, often yes. Relative to the *risk and capital* required to warehouse tail exposure, not necessarily.

Selling volatility is like collecting rent on a building in an earthquake zone. Most months, you'll feel brilliant. Your spreadsheet will look like it's been blessed. But you're short the earthquake. You're being paid for it every month, and the payment is real. So is the liability.

The right emotional posture is not "I found an arbitrage." It's "I'm running an insurance business." Insurance businesses can be fantastic. They also go bankrupt when they confuse premium income with safety.

A picture worth a thousand cautions

The relationship between implied and realized volatility is easiest to respect when you can see it. The green zones are the "normal" world short-vol traders dream about: implied above realized. The red zones are the world they fear: realized outruns implied and the insurance seller takes the hit.

You don't need the exact historical series to get the lesson. The pattern is the point: green dominates over long horizons, red arrives in clusters, and when red arrives it tends to arrive with attitude.

That is the bargain VRP offers you. The market will often pay you to sell uncertainty. It will occasionally demand a refund, with interest.

A practical way to measure VRP (without pretending it's a lab experiment)

If you're a trader, you want something operational. A clean, trader-friendly proxy is:

$$\mathrm{VRP}_{30\mathrm{d}} \approx \mathrm{VIX} - \text{Realized Vol}_{30\mathrm{d}}.$$

This is not the academically pure definition (variance is cleaner than volatility, and expectations matter), but it's a useful dashboard reading. It tells you whether the option market is charging a big premium over what has actually been happening.

When that spread is wide, option sellers are being paid more to take the other side. When the spread is tight or negative, the market is either complacent (tight spread) or in the middle of being punched (negative spread).

Two clarifications keep you honest:

- VRP is a *distribution*, not a constant. It wanders, and it changes regimes.
- A high VRP doesn't guarantee profits tomorrow. It suggests the market is offering a richer insurance premium *if* you can survive the period you're insuring.

Why equity index VRP is especially persistent

Equity indexes have a unique cocktail that makes VRP feel stubborn:

- Equity crashes are fast and correlated, so protection is valuable.
- Many institutions are long equities by mandate, so they are natural buyers of puts.
- Short-vol exposures can be capital intensive and politically dangerous inside an organization.

Even if you personally are brave, your investors might not be. Even if your model is sound, your drawdown tolerance might not be. So fewer players are truly natural, size-insensitive sellers of crash convexity. Less supply means higher prices, which means higher implied volatility relative to realized. That's VRP, expressed as market microeconomics.

The two-month trap: when VRP harvesters get overconfident

Most short-vol disasters have the same opening scene: a trader (or a fund) discovers VRP, sells options, collects premium, and then decides the premium is "income." They lever it. They smooth it. They start believing the strategy has a "Sharpe ratio" instead of a "cliff."

Then the regime changes. Volatility clusters. Correlations snap to one. Liquidity thins. Realized volatility outruns implied volatility. The premium they collected over months gets handed back in days.

This is not a moral story about greed. It's a mechanical story about negative convexity. When you're short volatility, the risk doesn't scale linearly with size. It scales like a bad rumor: slowly at first, then all at once.

Thinking like a volatility trader means respecting that VRP is compensation for providing liquidity and insurance *when it's least comfortable*. If you size the trade so you can't be there in the uncomfortable moments, you're not capturing VRP. You're renting it with a balloon payment.

Actionable takeaway

Treat VRP like an insurance business, not a hack. Before selling any premium, compute a simple 30-day spread: $\text{VIX} - \text{RV}_{30d}$ (or the closest equivalent on your underlying). If the spread is healthy, you're being paid to write insurance—but only take the trade if you can survive a fast move where realized volatility temporarily exceeds implied. Concretely: size so that a *doubling* of realized volatility over a short window (and a correlated IV spike) hurts, but doesn't force liquidation. If you can't size it that way, you don't have an edge; you have a time bomb with a coupon.

3.5 Vol of Vol: When Uncertainty Changes

Volatility has a personality disorder. One month it's a house cat: purring, predictable, content to nap in a sunbeam. The next month it's a feral animal that lives under the porch and bites anyone who gets close.

That instability isn't a poetic observation—it's a tradable fact. Volatility clusters. Calm tends to beget calm, chaos tends to beget chaos, and the shift between the two can be brutally discontinuous. The market doesn't gradually "turn up" risk like a dimmer switch—it often flips regimes like a breaker.

This is where a lot of otherwise sensible risk management goes to die. People size trades using yesterday's volatility as if they're setting cruise control. But volatility is not speed. It's the road surface. And the road can go from dry pavement to black ice in a single headline.

The meta-risk: uncertainty about uncertainty

You already have two useful mental buckets:

- Realized volatility: how violent the market *was*.
- Implied volatility: how expensive the market thinks future violence *might be*, after you account for supply/demand and insurance premiums.

Vol of vol adds a third bucket, and it's the one that makes markets feel haunted:

How unstable is volatility itself?

Sometimes you can live with a market that's noisy if it's consistently noisy. You can size down, widen stops, sell options with discipline, hedge with structure. The real problem is when volatility becomes unpredictable. That's when models break, correlations shift, hedges fail, and even "safe" positions start behaving like they drank too much espresso.

Vol of vol is the market's pricing of that instability. It's the "fear of fear": not just fear that prices will move, but fear that the *rules of movement* are changing in real time.

A day trader's version of the concept

You don't need an index or a PhD to feel vol of vol. If you've ever traded through a period where the S&P alternated between sleepy mornings and savage afternoons, you've felt it.

One week, a 0.7% intraday swing is a big deal. The next week, 0.7% is the warm-up. Your instincts become useless because your "normal" recalibrates daily.

In those periods, people start using phrases that sound emotional but are actually diagnostic:

- "The tape feels jumpy."
- "Liquidity is weird."
- "Everything's whipping around."

That's not just high volatility. That's high *volatility of volatility*: the amplitude and tempo of volatility are changing fast. The environment is unstable.

Why volatility clusters: the market is not a coin-flip machine

If returns were independent coin flips, volatility wouldn't cluster. It would be evenly noisy. But markets are social systems with feedback loops.

When the market drops, portfolios lose value. Risk managers cut exposure. Trend-followers sell. Dealers hedge short gamma by selling into weakness. Headlines get darker. Correlations rise. The system becomes more tightly coupled, like a bridge vibrating under load.

That coupling increases realized volatility, which increases implied volatility, which increases margin requirements, which forces more de-risking, which increases realized volatility again. Volatility clustering is partly psychology, partly plumbing.

The opposite happens in calm regimes. When markets drift upward and nothing is breaking, risk budgets expand. Vol sellers add size. Correlations look benign. Realized volatility stays low, which makes implied volatility feel expensive, which encourages more selling, which suppresses realized volatility further. Calm can self-reinforce, too.

So the market tends to settle into regimes:

- *Quiet regime:* low realized vol, low implied vol, tight spreads, generous liquidity.
- *Loud regime:* high realized vol, high implied vol, gappy moves, fragile liquidity.

The terrifying part is that regime changes often arrive before your dashboard catches up.

Why ``yesterday's VIX" is a dangerous position-sizing tool

It's tempting to treat VIX like a thermostat: if it's at 14, risk is "low"; if it's at 28, risk is "high." That's better than trading blind, but it's incomplete.

Two problems:

- VIX is a 30-day implied measure. Your risk might be intraday, overnight, or event-driven.

- The real risk is not the level of volatility. It's the *rate of change* of volatility—how quickly the environment can shift.

A VIX of 14 can be a trap if it's 14 after months of calm and positioning is crowded. A VIX of 28 can be a gift if it's 28 after panic and the market has already de-levered. The level tells you something. The stability of the level tells you more.

Vol of vol is the missing lens: it helps explain why two "VIX 18" environments can feel completely different. In one, volatility meanders gently and strategies behave. In the other, volatility is snapping around and everything becomes a coin toss.

Meet VVIX: implied volatility of VIX itself

The cleanest institutional expression of vol of vol is VVIX.

VVIX is commonly described as the "volatility of volatility" index: it's derived from prices of options on VIX, in a similar spirit to how VIX is derived from options on the S&P 500. If VIX is the market's price for equity index volatility, VVIX is the market's price for *how unstable that volatility might be.*

If that sounds abstract, ground it in the lived experience of a vol trader:

- VIX options pay off when VIX moves.
- VIX moves when equity volatility regimes change.
- So VIX options are insurance on the instability of the insurance market.

People buy them when they don't just fear a market move—they fear the volatility market itself whipping around, repricing violently, and breaking hedges.

VVIX tends to be jumpy, mean-reverting, and asymmetric: it often pops when equities sell off and the market starts worrying about a chain reaction. That's the "fear of fear" in traded form.

Vol of vol and the dealer's headache: when hedging becomes the event

In calm markets, the dealer's job is mostly an exercise in good housekeeping: manage delta, watch gamma, clip spreads, keep inventory under control. In unstable regimes, the job turns into triage.

When volatility is stable, you can hedge predictably. When volatility is unstable, hedging becomes path-dependent and expensive:

- Bid-ask spreads widen, so every hedge adjustment costs more.
- Liquidity becomes intermittent, so you can't always trade the size you want.
- Correlations shift, so your "diversified" hedges start moving together.
- Jumps make dynamic hedging less effective, because you can't hedge in the gap.

This is why vol-of-vol matters even if you never touch VIX products. It changes the *quality* of liquidity and the reliability of assumptions. It's the difference between a fast market and a brittle market.

A brittle market is the one that blows through strikes, triggers stop cascades, and prints those candles that look like someone dropped the chart.

The practical implication: your strategy's edge is regime-dependent

Almost every volatility strategy has a regime it loves and a regime that tries to kill it.

Short-vol carry tends to love stable, grinding markets where implied stays above realized and volatility mean-reverts politely. Long-vol convexity tends to bleed in those regimes but can pay explosively when the regime breaks.

Vol of vol is the regime detector that keeps you from confusing "my strategy has an edge" with "the market is currently feeding my strategy."

Even strategies that aren't explicitly "vol strategies" are exposed:

- Trend-following often performs differently when volatility is stable versus when it's whipsawing.
- Mean reversion strategies can get chopped to pieces when the intraday range expands erratically.
- Levered portfolios can go from fine to forced-selling when margin requirements jump with volatility.

You don't need to predict the next regime. You need to notice when regime risk is rising and stop pretending your backtest is a seatbelt.

A trader's way to think about regime shifts: ``quiet can be fragile''

There's a counterintuitive truth that good vol traders carry around like a superstition: *the calmest markets can be the most fragile.*

In long quiet stretches, investors sell more volatility, increase leverage, tighten stops, and generally behave as if the world has become safer. The system becomes optimized for calm. That optimization is exactly what makes it vulnerable. When volatility finally rises, the unwind is crowded.

This is why vol spikes can look like phase transitions. It isn't only "new information." It's also the market discovering that it built too much risk on the assumption of continued calm.

Vol of vol is the price of that fragility.

How to monitor vol of vol without fancy tools

If you have access to VVIX, it's a direct window into this risk. But even without it, you can build a practical proxy mindset.

Look for instability in the volatility complex:

- Is VIX whipping around day to day, not just trending?
- Are intraday ranges expanding unpredictably?
- Is implied volatility moving more than the underlying?

- Are you seeing frequent "vol shocks" around headlines that used to be ignored?

Also watch cross-asset tells. When vol of vol rises, correlations often rise too. Assets that "shouldn't" move together start moving together. That's the market's risk-off reflex becoming synchronized.

None of these signals are perfect. The goal isn't perfection. The goal is not being surprised by a regime change that was telegraphing itself through instability.

The risk management adjustment that matters most: cut optionality, not just size

When volatility becomes unstable, simply cutting position size is good but sometimes insufficient. The more powerful move is to reduce hidden *short convexity* exposures—positions that lose faster as markets move faster.

Short convexity can be obvious (short options) or sneaky (levered ETFs, tight stop-loss strategies in gappy markets, selling downside "because it's worked for years," certain yield-enhancement overlays). In a stable regime, these positions can look like disciplined income strategies. In an unstable regime, they can behave like a margin call generator.

If vol of vol is rising, the correct instinct is not "how do I keep earning carry?" It's "how do I avoid being forced into the worst trade of my life at the worst time?"

Actionable takeaway

Add a *regime check* to your routine that focuses on instability, not just level. Once a day (or once a week if you're slower-moving), record: (1) the current VIX level, (2) the *change* in VIX over the last 5 trading days, and (3) whether the last week's intraday ranges feel stable or erratic compared to the last month. If VIX is rising *and* its day-to-day moves are getting jumpier, treat that as a vol-of-vol warning: reduce positions with hidden short convexity, widen your risk buffers, and assume your hedges will behave worse than your model says. The goal isn't to predict chaos—it's to stop being sized like the calm will last forever.

Chapter 4

Taming the VIX: Futures, ETPs, and The Curve

If the S&P 500 is the heartbeat of the market, the VIX is its adrenaline level. Everyone watches the 'Fear Gauge,' but attempting to trade it without understanding the machinery behind it is the fastest way to blow up a portfolio. Here, we strip away the academic jargon to reveal what the VIX actually measures, why you can't buy it directly, and why the products that track it-like futures and ETPs-often behave like a hall of mirrors. We will move from the math of the index to the geometry of the curve, ensuring you know the difference between buying insurance and burning money.

4.1 What the VIX Actually Is

If you've ever watched the market on a stressful day, you've seen the VIX flashed like a vital sign: green arrows, red arrows, breathless narration. When it pops, commentators reach for the same nickname-*the Fear Gauge*. It's not wrong in the way calling a thermometer a "fever stick" isn't wrong. Fever and fear show up in the same situations. But a thermometer isn't *made* of panic; it's made of physics. The VIX isn't made of emotion either. It's made of option prices-thousands of tiny insurance quotes-distilled into a single number that answers a very specific question: *What level of volatility is the S&P 500 options market implying over the next 30 days?*

That one sentence is where most traders either get clarity—or step into a swamp of misconceptions. The VIX is not "how scared people are," full stop. It's not a mood ring for Wall Street. It is *implied volatility*, annualized, for a roughly one-month horizon. In plain English: it's the market's best guess of how violently the S&P 500 might swing, backed by real money bidding for protection and leverage. And because it's a guess embedded in prices, it can be wrong, early, exaggerated, muted, or oddly calm in situations that *feel* terrifying. That's not a flaw. That's the point. It's a forward-looking price.

Here's the mental reframe that makes the VIX click: imagine you walk into an insurance marketplace where you can buy protection against the S&P 500 having a rough month. The stall owners aren't therapists; they don't sell feelings. They sell contracts. If the crowd suddenly wants more protection, the quotes go up. If the crowd is relaxed, or if there's plenty of sellers competing, the quotes go down. The VIX is essentially a clean, standardized way of summarizing those quotes into a single implied-volatility number. Calling it fear is like calling homeowners insurance "hurricane vibes." Sometimes correlated, never identical.

A second misconception is even more important: many people think the VIX measures what *just happened*. That's realized volatility—the bumpy road in the rearview mirror. The VIX is about the road ahead, or at least what drivers are willing to pay to insure it. That difference explains a ton of "why is the VIX doing that?" moments. If the S&P 500 drops 2% today, realized volatility just rose—something happened. But the VIX only rises if option prices imply *more* turbulence over the next month than they implied yesterday. Those often coincide, but not always. The VIX can rise on an up day if traders think the rally is fragile, or if there's an event looming (an inflation print, an election, an earnings cluster) and protection gets bid anyway. Conversely, the VIX can fall on a down day if the selloff is orderly, the event risk just passed, or dealers are aggressively selling options and forcing implied volatility down.

A quick way to feel the difference: realized volatility is your bank statement-what your portfolio actually did. Implied volatility is the insurance premium you're quoted going forward. If your car got into an accident last week, your insurance company doesn't automatically assume you'll crash tomorrow. But if a hailstorm is forecast for next week, premiums can jump even if the sun is shining today. VIX is much closer to the hailstorm premium than the accident report.

So what does "VIX = 20" *mean*? It doesn't mean "20% chance of a crash," and it doesn't mean "stocks will drop 20%." It's an annualized volatility number. Volatility, in this context, is the statistical idea of typical movement-how wide the distribution of returns is expected to be. Roughly, traders often translate VIX into a monthly "expected move" by dividing by $\sqrt{12}$ (because there are about 12 months in a year). So if the VIX is 20, the market is implying something like a 20% annualized standard deviation, which is about $20/\sqrt{12} \approx 5.8\%$ as a one-standard-deviation move over the next month. That's not a guarantee; it's a way to convert a single index number into a gut-level sense of the options market's pricing.

Notice what's hidden in that conversion: *probabilities*. VIX isn't predicting a single outcome, like "SPX will be down 3% next month." It's summarizing how expensive a whole range of outcomes has become. The options market isn't betting on one horse; it's pricing an entire racetrack-small moves, big moves, disasters, melt-ups, all of it. When that whole racetrack becomes more expensive, implied volatility rises, and the VIX rises with it.

Now for the subtle, beautiful weirdness: the VIX is not calculated from a single option. It's derived from a *strip* of S&P 500 index options-out-of-the-money puts and calls across many strikes-using quoted bid/ask prices. Think of it as a crowd-sourced volatility estimate, with each strike contributing a vote weighted by how it helps represent variance over the next 30 days. That structure is why the VIX can respond so violently during stress: deep out-of-the-money puts—the "tail insurance" strikes that barely trade in calm times—can suddenly get repriced dramatically when people panic-buy protection. Even if the S&P 500 itself hasn't moved that much yet, the market can start pricing a fatter left tail. The VIX listens to that tail.

It also explains why the VIX can jump in a way that feels disconnected from reality: it's based on *quotes*, not your personal sense of fairness. On normal days, bid/ask spreads are tight, and quotes track trades closely. On chaotic days, spreads widen, market makers protect themselves, and the quotes can move fast. The VIX is a number that takes those quotes seriously—because those are the prices you'd actually have to pay (or accept) to transact. The index is not trying to be comforting; it's trying to be tradable truth.

Another core point: VIX targets a *constant maturity* of about 30 days. That means it isn't the implied volatility of "the option that expires this month." It's designed to stay anchored to roughly one month, every day, by pulling from options in a maturity window and interpolating to a 30-day measure. This is why VIX can behave smoothly through calendar time even as individual option expirations roll off. It's always trying to answer the same question-what's implied over the next 30 days-rather than letting the question drift to "next Friday" and "two Fridays after that."

At this point, most people ask the practical question: *Can I buy it?* Not directly. The VIX is an index-a computed number-like an exchange rate you see on a screen. You can't reach into the screen and own it. You can trade *derivatives* tied to it (futures and options), and you can buy products that hold those derivatives (ETPs), but the index itself is a reference point. This is where many traders get hurt: they look at spot VIX, buy a VIX-linked product, and assume they've bought "the thing they saw on TV." In reality they've bought an instrument linked to a *different* thing (often a basket of futures), with its own physics. The distance between "VIX the number" and "VIX the trade" is where the curve, carry, and decay live.

Before getting into any instruments, it's worth confronting one more false intuition: people assume the VIX is "the opposite" of the S&P 500, like a perfect mirror. In practice the VIX is often *negatively correlated* with stocks-markets drop, implied volatility rises-but it isn't a simple inverse ETF. The relationship is lopsided because options are lopsided. Investors tend to buy puts for protection more urgently than they buy calls for upside, especially in institutional size. Dealers who sell those puts hedge dynamically, and that hedging can amplify moves and reprice implied volatility. That's why VIX spikes can be sharp and vertical, while declines in VIX often look like slow leaks. Fear tends to arrive by elevator and leave by stairs, but the mechanism isn't poetry—it's supply, demand, and hedging pressure.

Here's a scenario that makes the "VIX can rise while stocks rise" idea feel intuitive instead of heretical. Suppose the S&P 500 is up 0.8% today, a nice green day. But the rally is concentrated in a few mega-cap names, breadth is weak, and tomorrow there's a CPI release that has a history of causing gap moves. Portfolio managers may decide: "I'll take the up day, but I want protection into tomorrow." They bid up puts and maybe sell some calls to finance them. Option prices reflect that demand. Implied volatility rises. The VIX rises. Everyone watching a green tape gets confused, but the insurance market is simply doing its job: pricing tomorrow's uncertainty, not celebrating today's return.

Flip it. The S&P 500 can fall and the VIX can fall too. Picture a market that has been bracing for a scary event—say a central bank decision. Implied volatility gets bid up in advance. When the decision hits, the market drops 1.2%, but the message is clear, liquidity holds, and traders can finally exhale because the unknown is now known. Protection is sold back. Implied volatility compresses. VIX drops. The day was red, but uncertainty got cheaper.

That brings us to the most tradable personality trait of the VIX: *mean reversion*. Stocks can trend for years. The VIX doesn't trend in the same way because volatility is not a company that can compound earnings. Volatility is a condition. It flares up, then it tends to cool down. That doesn't mean it's safe to short volatility whenever it's high-high can go higher, and fast-but it does mean VIX spikes are historically more like storms than like permanent climate. The air pressure can drop suddenly, but it usually doesn't stay at hurricane levels forever.

Mean reversion in volatility also makes intuitive sense when you remember what option prices represent: insurance premiums. In a panic, people overpay for insurance because they *need* it now and sellers demand a premium for taking the other side. As the panic fades, the urgency fades, sellers return, competition increases, and premiums normalize. Even without a "news catalyst," time itself can be a solvent: as days pass without disaster, the probability of disaster over the remaining 30-day window shrinks. Less uncertainty, cheaper options, lower implied volatility, lower VIX. Not magic—just the calendar doing what calendars do.

The VIX also carries a kind of "gravity" because of how many strategies lean against it. Institutions sell options for yield. Dealers hedge in ways that stabilize or destabilize depending on positioning. Systematic funds adjust exposure when realized volatility changes. All of that creates feedback loops where volatility is both a reflection of risk *and* an input to behavior. That's why VIX is such a powerful index: it isn't only measuring uncertainty; it can become part of the machinery that shapes market dynamics.

One more practical translation helps: VIX is often described as "30-day implied volatility," but what it really captures is *implied variance* extracted from a wide strip of option prices and then converted back into a volatility number. That nuance matters because variance is what adds linearly across time and across components of a portfolio. The VIX methodology is built to be model-free in spirit: it doesn't assume a particular option pricing model like Black-Scholes with a single volatility parameter. Instead, it takes the market's surface-many strikes, many quotes-and backs out a single "expected variance" number for the horizon. When you know that, VIX feels less like a mystical fear meter and more like a robust accounting identity: a way to read the options market's balance sheet of risk.

When traders treat VIX as a mood, they make mood-based trades: "Everyone's scared, so I'll buy VIX." When traders treat VIX as a price, they start asking better questions:

- What is the market pricing for the next 30 days, and how does that compare to what I think is likely?
- Is implied volatility expensive or cheap relative to realized volatility?
- Is there an event that justifies a premium, or is the premium leftover adrenaline?

Those questions are the difference between gambling on a headline and trading a structure.

A final caution that saves real money: the VIX is *not* your P&L. You can be "right" about the VIX and lose on the instrument you chose. Spot VIX might spike, but the VIX future you're exposed to might move less. Spot VIX might mean-revert, but your product might decay faster than you expected. The VIX itself is a measurement. The trade is a machine built around that measurement, with gears like term structure and rolling.

Knowing what the VIX is-exactly-is how you stop blaming the machine for doing what it was designed to do.

Actionable takeaway: Before you place any VIX-related trade, write down (literally, on paper) a single sentence answering: "I am expressing a view on __________ over the next _____ days." If your sentence says "fear" or "today," you're still trading a nickname. Rewrite it until it says "30-day implied volatility of the S&P 500 as priced by SPX options." Then ask one more: "Am I trading the index, a futures month, or an ETP basket?" If you can't name which, don't click.

4.2 The Term Structure: Contango and Backwardation

The VIX number you see on a screen is a *quote about the next month*. Traders, unfortunately, don't get to trade quotes. They trade contracts. And the moment you move from "the VIX" to "a VIX future," you discover the market's best practical joke: there isn't one VIX price—there's a *menu* of VIX prices, each tied to a different expiration month. That menu is called the *term structure* (or "the curve"), and it behaves less like a stock chart and more like an insurance rack at the store: same product category, different price tags depending on how long you want coverage.

Think about buying storm insurance for your house. Coverage for "this weekend" is priced differently than coverage for "the entire next year," even though they're both insurance against weather. Volatility works the same way. A one-month VIX future is the price of near-term uncertainty. A six-month VIX future is the price of uncertainty with time to simmer, spread out, and mean-revert. Put those futures next to each other—from soonest expiration to farthest—and you get a curve that quietly dictates whether volatility trades feel like pushing a boulder uphill or sliding downhill with a tailwind.

This curve matters for one simple reason: time is not neutral in volatility. In stocks, time tends to reward you (dividends, growth, drift). In volatility, time is often a toll booth. If you're long volatility exposure via futures, you're frequently paying for the privilege of holding it. If you're short, you're frequently collecting the premium. The curve is where that toll is posted.

One concept, many prices

A VIX futures contract is a standardized bet on where the VIX will be at a specific settlement date. The front-month contract (the one expiring soonest) is the market's best guess of where 30-day implied volatility will be around that expiration. The second-month contract is a guess for a later date, and so on. Line them up, and you'll usually see something like:

- near-term futures clustered closer to today's "spot" VIX level (because time to expiration is short), and
- longer-dated futures priced with a different kind of wisdom: not "what's happening right now," but "what's likely after the dust settles."

That's the heart of term structure: it's the market pricing volatility across time horizons. It's also the first place retail traders realize the VIX ecosystem is not a single lever. You can be "right" that fear will rise, and still lose money if you chose the wrong month, at the wrong point on the curve, in the wrong regime. Volatility is a timing instrument, and the term structure is the clock.

Contango: the curve of calm

In "normal" markets, the VIX futures curve tends to slope upward. Near-term futures trade cheaper than longer-term futures. That upward slope is called *contango*. If you're new to futures, contango sounds like a fancy word for a fancy reason; in practice it's a plain message from the market: *near-term stress is relatively cheap, long-term uncertainty is priced higher.*

Why would longer-dated VIX futures be higher in quiet times? Because volatility mean-reverts, but it doesn't mean-revert to zero. It mean-reverts to a *positive* level, and markets know that "calm" is often temporary. When spot VIX is low, the market tends to expect it will drift back up toward a more average level over time. So longer-dated contracts sit above spot like a gentle reminder: "Sure, today is quiet. Don't assume quiet is forever."

There's also an insurance-market intuition: sellers of volatility (the people effectively writing insurance) usually demand a premium for tying up risk for longer periods. One-week protection is one kind of promise. Six-month protection is a different kind of promise. In calm conditions, it's rational that the longer promise costs more.

For traders, contango is not just a shape. It's a gravitational field. It's the regime where long volatility carry tends to be negative, and short volatility carry tends to be positive. That's not a moral statement; it's physics.

Backwardation: the curve of panic

Then there's the other shape—rarer, sharper, and emotionally louder. In a real scare, near-term VIX futures can trade *higher* than longer-dated futures. That downward slope is *backwardation*. It's the curve that says: *the house is on fire right now, and everyone wants the fire extinguisher today.*

Backwardation is what happens when immediate protection becomes priceless. The market isn't paying up for "uncertainty in general." It's paying up for a very specific thing: near-term tail risk. Traders don't care what volatility might be in five months if they're worried about what it might be in five hours. That urgency lifts the front of the curve.

And here's the twist that surprises people: backwardation can coexist with the belief that things will eventually calm down. In fact, backwardation often *requires* that belief. The curve is screaming "now is terrifying," while also whispering "this won't last forever." That combination—acute fear now, expected normalization later—is exactly what a market panic looks like.

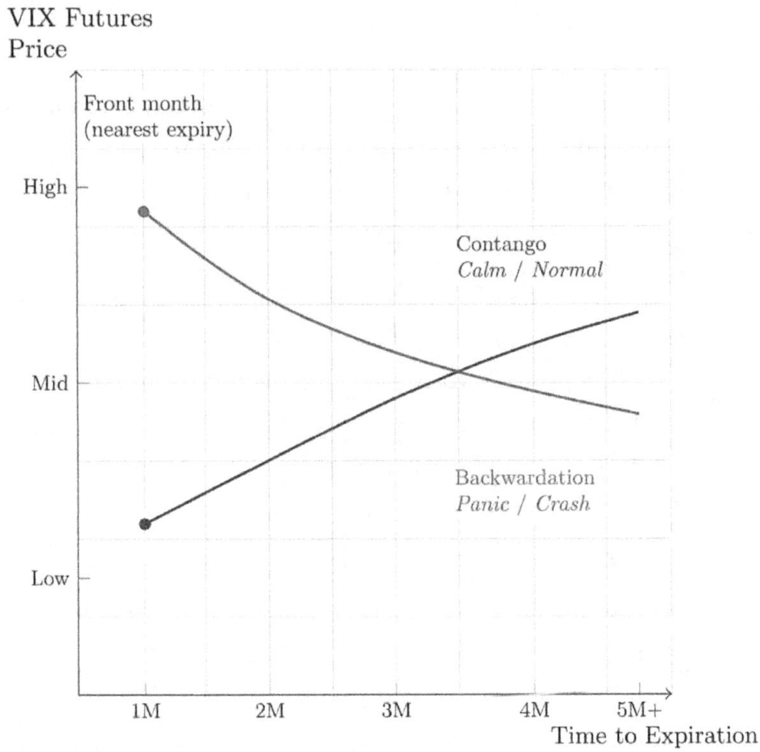

The curve is the ``price of time"

If you only remember one sentence about VIX term structure, make it this: *the curve tells you what the market charges for moving your volatility exposure through time.*

That's not philosophy—it's the practical reality of maintaining a VIX futures position. A futures contract expires. If you want to keep your exposure, you eventually have to "roll" from the expiring contract into a later one. Rolling means selling what you own and buying the next contract. In an upward-sloping curve (contango), the next contract is typically more expensive than the one you're selling. That difference is a headwind. In a downward-sloping curve (backwardation), the next contract is cheaper. That difference can be a tailwind.

This is where traders get blindsided: they focus on the level of VIX and ignore the slope of the curve, and the slope quietly becomes their dominant return driver. It's the difference between buying insurance occasionally and *running an insurance subscription*. If you're long front VIX futures in contango, you're paying a recurring premium. If you're short them, you're often collecting it—until the day the premium becomes a claim.

A story the curve tells about expectations

The term structure also contains a narrative about what the market expects *after* the current headlines fade. Consider two very different days:

Day A: Spot VIX is 14, the front-month future is 15, and six months out is 18. That curve says: "Today is calm, but calm is fragile; long-run volatility is priced higher than today's print." Traders are not panicking. They're pricing the world as uncertain in the normal way, not the emergency way.

Day B: Spot VIX is 35, the front-month future is 33, and six months out is 26. That curve says: "The next few weeks are chaos, but months from now we expect the fire to be mostly out." It's not optimism about stocks; it's mean reversion being priced into volatility.

That second day is also a clue to why volatility spikes can be so dramatic yet so temporary. When the curve is in backwardation, the market is already pricing in some cooling. If conditions stabilize even slightly, the front of the curve can fall fast—not because everyone is suddenly brave, but because the near-term insurance premium collapses once the forced buyers stop forcing.

Why contango is the default, and why that's dangerous

Most of the time, markets are not in crisis. That's almost a tautology, but traders forget it whenever the VIX hits the headlines. In most "normal" windows, the VIX curve tends to live in contango because the world tends to live in "not-a-panic." The practical implication is uncomfortable: many long volatility products and strategies are structurally fighting the default shape of the curve.

This is not the same thing as saying "being long volatility is stupid." Long volatility can be profoundly intelligent when you have a reason—an event risk you understand, a tail hedge you can afford, a tactical trade with defined downside. The mistake is treating long volatility like a long stock position: buy it, hold it, and wait for it to work out. In contango, waiting is expensive. The curve is literally charging you rent.

Meanwhile, short volatility strategies—explicit or disguised—often look brilliant in contango. They clip premium. They harvest carry. They feel like free money with a good Sharpe ratio. Until they don't. Backwardation is the market's way of saying the insurance company is paying out, and suddenly the "rent" flips directions. Anyone leaning on contango as a permanent law of nature discovers that the curve has teeth.

Backwardation isn't a crystal ball

It's tempting to treat backwardation like a prophecy: "The curve is inverted, so stocks must keep falling." That's not how it works. Backwardation is primarily a statement about *immediacy*. It's the market charging a premium for near-term protection. Sometimes that happens at the start of a prolonged crisis. Sometimes it happens near the middle, when the news is ugly but the worst is already in motion. Sometimes it happens for technical reasons—positioning, hedging flows, liquidity strains—that unwind faster than the headlines.

So the term structure is less like a fortune teller and more like a live auction board. It tells you what people are paying *right now* for volatility at different horizons. Your edge, if you have one, comes from identifying when that auction is mispricing the future relative to your thesis and risk tolerance.

Practical reading: three curve questions that actually help

When you pull up a VIX futures curve (most platforms show the monthly expirations as a line chart), the useful questions are not "Is it up?" or "Is it down?" The useful questions are:

1. *What's the slope at the front end?* The front two contracts are where carry gets loud. A steep upward slope is a tax on long exposure; a steep downward slope can fuel it.
2. *How far is spot from the front-month future?* When spot VIX is far above futures, the market is often implying that today's stress won't persist into settlement. That gap can close painfully fast if you're long the wrong thing.

3. *Is the curve "humped" or smooth?* Sometimes mid-curve contracts price higher than both near and far. That's the market pinpointing event risk in a particular window—like an election month—rather than pricing generic fear.

Those questions aren't academic. They're operational. They tell you whether time is likely to pay you or charge you for holding your position, and they hint at which contract month is truly expressing your view.

A trader's analogy: the curve as a set of thermostats

Spot VIX is the temperature at the window right now—drafty, noisy, reactive. The futures curve is a set of thermostats set for different dates. The front-month thermostat can change quickly; it reacts to immediate drafts. The six-month thermostat is steadier; it's anchored to the building's long-term climate. When markets panic, the window temperature spikes and the near thermostat jumps too, but the far thermostat may move less because it assumes the building won't stay on fire forever.

If you're trading volatility, you're not just guessing today's temperature. You're choosing which thermostat you're betting on. That choice is the difference between a contract that moves with every headline and one that barely flinches unless the crisis looks durable.

Where retail traders step on rakes

The most common retail mistake is treating the curve like background decoration. A trader sees the VIX spike on a scary day, buys a VIX-linked product, and expects a clean inverse relationship to the S&P 500. But the product is often exposed to a *rolling* position in the front of the futures curve, and the curve's slope can dominate outcomes.

Here's what that feels like in real life: the market sells off, the VIX pops, you buy "volatility." The next day the market chops sideways, the VIX drifts lower, and your product bleeds more than you expected. You weren't necessarily wrong about risk; you were long a position that pays rent every day in contango, and you didn't stay long enough (or get the right magnitude) to overcome that rent. The curve didn't care about your narrative. It charged you anyway.

On the other side, short-volatility trades can feel like printing money in contango—small gains that arrive with metronomic regularity. The curve can make you overconfident. You start believing you've discovered a stable income stream. Then a real stress hits, contango collapses into backwardation, and the same mechanism that used to drip gains starts spraying losses. Volatility is one of the few places in markets where the "steamroller" metaphor is overused and still underestimates the problem.

Actionable takeaway: never trade VIX exposure without a curve check

Actionable takeaway: Before putting on any VIX futures, VIX options, or VIX-linked ETP trade, spend 30 seconds doing a "curve check." Pull up the front six VIX futures and write down just two numbers: the front-month price and the second-month price. If the second month is higher, you're in contango at the front end and time is likely *charging* long exposure. If the second month is lower, you're in backwardation at the front end and time may be *paying* long exposure (while warning you the market is under real stress). Then force a concrete decision: either (a) size your long-vol idea small enough that you can afford the carry, or (b) shorten your holding period and define your exit in advance. The curve isn't a detail—it's the price tag on time.

4.3 The Roll: Understanding Carry

A lot of volatility trades don't die in a blaze of glory. They die like a phone battery: 1% here, 0.7% there, a slow drain you barely notice until you're stranded. That drain has a name—*carry*—and in the VIX futures world, carry is mostly about one mundane, unavoidable chore: *the roll.*

A VIX future has an expiration date. If you want "long volatility" exposure that lasts longer than a single contract's life, you have to keep swapping into the next one. That sounds harmless—like renewing a subscription. But when the VIX futures curve is upward sloping (contango), renewal doesn't come at the same price. You repeatedly sell something cheaper and buy something more expensive. You don't need a market crash to lose money. You can lose money simply by staying alive.

The simplest version of the roll

Picture a trader who wants ongoing exposure to the front of the VIX futures curve—because that's where volatility tends to respond the most when equities get hit. They buy the front-month VIX future today. A week passes. The contract is now one week closer to expiration. Another week passes. Closer still. Eventually it expires, and they must replace it with the next month.

That replacement is the roll. Mechanically, it's just:

- Sell the contract that's about to expire (the "near" contract).
- Buy the next contract out (the "far" contract).

The roll itself is not an opinion. It is plumbing. It happens whether you're bullish, bearish, or emotionally exhausted. And in contango, that plumbing leaks.

Convergence: the quiet force that moves futures

The roll becomes painful because VIX futures don't float freely in space. As expiration approaches, a VIX future tends to *converge* toward the value it will settle to at expiration (often described as converging toward "spot," though settlement mechanics have their own quirks). Convergence is not a vibe; it's the definition of a futures contract. If a future didn't converge, it would be a permanent arbitrage invitation.

Here's the important part: in contango, the front-month future often starts *above* the current spot VIX level. As time passes, it often "slides down" toward settlement. If you're long that contract, this slide is headwind even if the VIX index itself goes nowhere.

That's why the roll is frequently called "structural decay" for long volatility products. The curve sets the initial premium; convergence slowly collects it.

Carry: the P&L you earn (or lose) for doing nothing

Carry is one of those trader words that sounds mystical until you translate it: *carry is what your position tends to earn or lose if the world stays roughly the same.*

In equity investing, "the world stays roughly the same" can still mean you make money because stocks have long-run drift. In VIX futures, "the world stays roughly the same" often means the curve gently bleeds you because contango keeps charging rent.

The carry of a VIX futures position depends heavily on the shape of the term structure. When the curve is in contango, long front-month exposure tends to have *negative carry* (a drag). Short front-month exposure tends to have *positive carry* (a tailwind). When the curve flips into backwardation, the sign can flip too—suddenly long volatility can have positive carry, and short volatility can feel like holding a lit match in a fireworks store.

The uncomfortable truth: many traders spend all their energy forecasting the next shock and almost none measuring the carry they're paying while they wait.

A concrete ``buy high, sell low" example

Numbers make this real fast. Suppose:

- Spot VIX is 15.
- The front-month future (F1) is 16.
- The second-month future (F2) is 17.

That's contango. Now imagine nothing dramatic happens for a couple weeks—equities drift, headlines rotate, but no real fear. As the front-month approaches expiration, F1 tends to sag toward where it will settle. Maybe it drifts from 16 down to 15.3. The second-month might drift too, but typically it becomes the new "front" and begins its own convergence process.

If you're long F1, you can lose 0.7 VIX points without the VIX index "doing anything." With a VIX futures multiplier of $1,000 per point, that's about $700 per contract—simply from time passing in the wrong curve regime.

Then comes roll day. You sell the now-lower front-month (say, 15.3) and buy the next month (say, 16.6). You just paid 1.3 points to keep your exposure. That 1.3 points is not "bad luck." It's the curve charging you for carrying a position that tends to pay off explosively in rare moments.

A short seller lives on the other side of that transaction. They sell the richer contract, buy it back later as it converges, and repeat. That repeatability is why short vol can look like a license to print money—until a regime shift turns "repeatable" into "catastrophic."

Negative roll yield: the silent killer

The drag created by selling cheaper and buying more expensive is commonly described as *negative roll yield*. "Yield" sounds polite; don't let it fool you. In contango, negative roll yield is a mechanical "buy high, sell low" loop embedded into the act of maintaining long volatility exposure.

This is also why many VIX-linked products have charts that look like melting ice. People interpret the decay as "volatility fell." Sometimes that's true. Often it's not the whole story. The more honest explanation is: *the product is paying the insurance premium over and over, and premiums add up.*

The best way to think about negative roll yield is as a subscription fee for convexity. Long front-end volatility is one of the most convex exposures available to liquid traders—it can explode upward during equity shocks. The market doesn't give convexity away for free. In contango, the fee is charged continuously through carry.

The stepwise slide: what it looks like in your P&L

There's a particular pattern long volatility traders learn to recognize, the way a sailor recognizes the smell of a storm. Your P&L bleeds in a staircase: small drops most days, occasional bigger drops around rolls, interrupted by sudden spikes when volatility finally wakes up. If the spike doesn't come soon enough—or isn't big enough—the staircase wins.

That staircase is not psychological. It's geometry: the curve is above spot, and each contract moves down the curve toward settlement. Then you jump up to the next contract (at a higher price) and start sliding again.

The roll dynamic is easiest to see as a stepwise path along an upward-sloping curve:

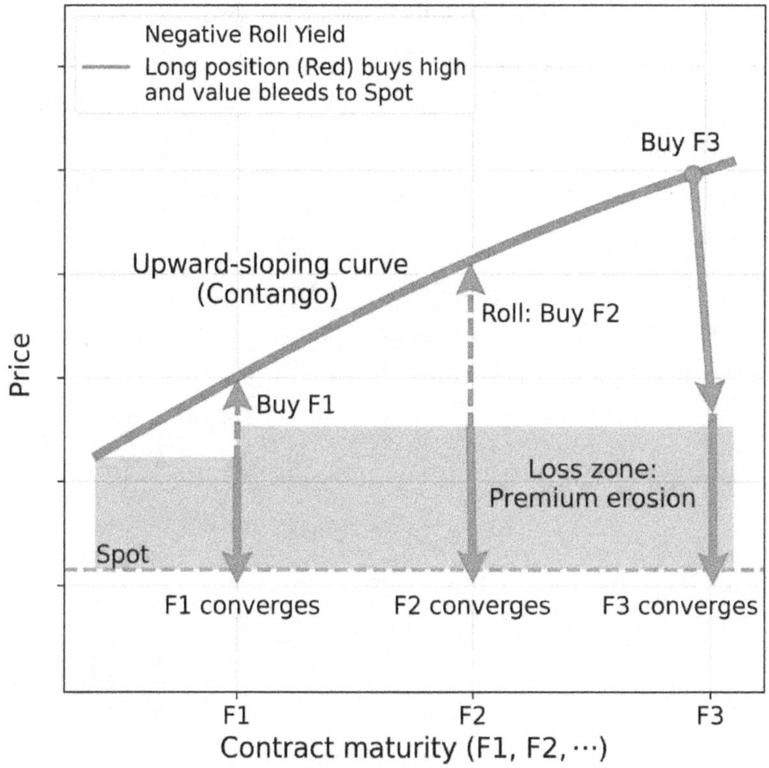

Roll yield isn't ``a fee," it's the market paying itself for disaster insurance

It's tempting to think of negative roll yield as an unfair fee imposed by product designers or brokers. In reality, it's the market pricing an uncomfortable truth: people are willing to overpay for protection, on average, because protection is most valuable when your brain is least capable of calm math.

That's not an insult; it's a human fact. When markets are quiet, investors still want some insurance in the drawer. When markets get loud, they want it *now*. The willingness to pay up for convexity is what makes selling volatility profitable over long spans, and it's what makes buying volatility expensive most of the time. The roll is simply how that long-run premium becomes a daily P&L.

If you've ever wondered why "long VIX ETP" charts seem to decay even across periods where the VIX doesn't look dramatically lower, the roll is the missing mechanism. A product can be roughly right on direction and still hemorrhage due to carry.

A quick table: separating the two kinds of ``loss''

One reason traders stay confused is they mix two losses into one emotional blob. There's the loss from being wrong about volatility *level* (VIX falls). And there's the loss from carry (the curve bleeds you even if volatility is flat). Keeping them separate makes you a better risk manager.

Source of P&L	*What it feels like*	*What's actually happening*
Volatility move (direction)	"VIX went down and I lost."	The futures price moved lower because implied vol expectations fell (or the contract repriced).
Carry / roll yield	"Nothing happened and I lost anyway."	In contango, the contract you own tends to converge downward, and rolling requires buying the next contract at a higher price.

Why short volatility ``earns carry" (and why that's not a free lunch)

If contango bleeds longs, it rewards shorts—at least in the same quiet regimes where the curve is upward sloping. A short seller can sell a richer front-end future and buy it back later after convergence has pulled it down. Then they roll by shorting the next rich contract. They're effectively running the opposite staircase: small gains that stack, punctuated by occasional sharp losses when volatility spikes.

That payoff shape is why short volatility strategies often look like disciplined income—until the moment they don't. The carry you "earn" is the premium for being the insurance writer. The premium exists because occasionally you have to pay claims, and those claims arrive during the exact moments when liquidity is thin and your risk limits feel tight.

So the roll is not just a long-vol problem. It's also the seduction mechanism for short vol: it pays you to keep the trade on. It encourages complacency. It makes you think you're being compensated for skill when you're being compensated for bearing tail risk. The market does not hand out Sharpe ratios for bravery; it hands them out for selling disaster insurance in calm weather.

Carry-aware thinking: the difference between a hedge and a hobby

The roll becomes manageable once you stop treating long volatility as a permanent posture and start treating it like a tool with a carrying cost. There are two carry-aware questions that change everything:

1. *How long can I afford to be right late?* If your thesis needs three months to play out, but the curve is steep contango, your position needs either very small size or a structure that limits carry bleed.
2. *What magnitude of volatility move do I need to break even on carry?* If you're paying, say, 0.8 VIX points per month in roll costs (conceptually), you need a volatility spike big enough to overcome that drag within your holding period.

This is where many "I bought volatility as a hedge" decisions fall apart. A hedge is not a bumper sticker. A hedge is a budget. If you can't articulate the budget—the carry you expect to pay—then the "hedge" is just an anxiety trade dressed up in grown-up language.

A practical mental model: you're renting convexity

Convexity is the quality that makes gains accelerate when the world gets ugly. It's why some volatility positions can save a portfolio during an equity air pocket. Convexity is also expensive, because it's scarce and most valuable in the moments everyone wants it simultaneously.

When you hold long front-end VIX futures exposure in contango, you are *renting* convexity. Rent is due every day. If the disaster doesn't show up, you don't get a refund for being cautious. You just paid for peace of mind and didn't need it. That's what insurance is supposed to feel like.

This mental model stops the self-sabotage cycle where traders keep doubling down on a long-vol position because "it's down so much it has to work." A rental doesn't "have to work." It just keeps charging you until you return the keys.

Actionable takeaway: calculate your carry before you fall in love with the trade

Actionable takeaway: Before holding any VIX-futures-based position overnight (directly or through an ETP), do a two-minute carry estimate. Write down the front-month future (F1) and second-month future (F2). Compute the simple spread $F2 -$

$F1$. If it's positive, you're in front-end contango and long exposure is paying rent. Then set a rule: either (a) define a maximum hold time (for example, "I will not hold this longer than 10 trading days unless volatility is actively expanding"), or (b) size the position so that losing roughly that spread over a month would be emotionally and financially boring. If you can't make the expected carry loss boring, you don't have a hedge-you have a stress amplifier.

4.4 The Widowmakers: Volatility ETPs and Decay

The most expensive sentence in volatility trading is: *"I'll just hold it until it comes back."* It sounds patient. It sounds disciplined. It's also the exact mindset that turns certain volatility ETPs into wealth composters. These products don't merely *fluctuate*. Many are engineered around a rolling VIX futures position that, in normal markets, behaves like a bar of soap in the shower: the longer you hold it, the less of it you have.

The danger is that they look familiar. They trade like stocks. They sit in the same brokerage account. They have tickers you can buy in one click. That familiarity is camouflage. A long-volatility ETP is not "a stock that tracks the VIX." It's usually a rules-based machine that owns VIX futures and rolls them forward every day, paying (or collecting) whatever the curve demands. If you don't respect the machinery, the ticker will educate you with real dollars.

The wrapper matters: ETP is a container, not a strategy

"ETP" is a broad label—exchange-traded product. In the VIX world, the big split is often between ETFs and ETNs. The distinction sounds like legal trivia until you realize one of them is essentially a promise.

An *ETN* (exchange-traded note) is unsecured debt issued by a bank. You're not owning a pile of futures in a vault; you're owning an IOU whose payoff is linked to an index. That means you inherit issuer credit risk, and the note can behave strangely when supply/demand for shares gets out of sync—trading at a premium or discount to its indicative value. Under stress, when you least want extra moving parts, you can discover that your "VIX hedge" also has "bank credit exposure" and "market plumbing exposure." None of that is theoretical. It's written into the product structure.

An ETF-style product can have its own complexities, but the key message is the same: the wrapper doesn't remove the roll. It industrializes it. The product *must* maintain exposure according to a stated methodology, which usually means buying and selling VIX futures on a schedule whether the trade is good or bad that day. A human trader can choose to stop paying rent. The product can't. It's a machine with no discretion and no shame.

What these products actually hold (and why that's the trap)

Most popular long-volatility ETPs don't hold "the VIX." They hold exposure to a *short-term VIX futures index*. In plain English: they own a blend of the front-month and second-month

VIX futures, and they rebalance that blend day by day so the portfolio always stays near a constant one-month maturity. That design makes the product responsive to near-term volatility shocks—which is what people want—but it also maximizes exposure to the part of the curve where contango tends to be steepest and carry tends to be most punishing.

This is where the trap snaps shut. A trader sees "VIX up" on a scary day and buys a long-vol ETP as a hedge. If the panic continues, the trade can work spectacularly. If the panic fades, the product often doesn't merely drift lower—it can resume its normal habit of bleeding through carry. And if nothing happens for long enough, the product can lose a shocking percentage of its value even if the VIX index spends years oscillating in a relatively stable range.

That's not because the product is "bad." It's because it is *honestly* doing what it said: maintaining short-term futures exposure in a market where short-term volatility insurance usually has a premium.

The two forms of decay: curve decay and path decay

Long-volatility ETPs have two separate enemies. Most traders meet the first and never recognize the second.

Enemy #1: Curve decay (carry). When the curve is in contango, the product is constantly rolling up the curve—selling cheaper futures and buying more expensive ones—and then watching those futures converge downward. That is negative roll yield, industrialized. In a quiet market, the product can lose value steadily simply because time passes and the curve keeps charging for exposure.

Enemy #2: Path dependency (compounding). Many volatility products are leveraged or reset exposure daily. Even without contango, daily compounding can erode returns when the underlying is choppy. You can be directionally "not wrong" and still lose money because the path—up and down and up and down—creates volatility drag. This is the same reason a leveraged ETF can underperform your naïve expectation over time. With volatility ETPs, you can get a double hit: the curve drags you, and the daily path rubs salt into the drag.

The brutal thing about path dependency is that it feels like the market is mocking you. You can watch a product spike on Monday, crash on Tuesday, spike again on Wednesday, and you'll swear it's "going nowhere." But your account balance has gone somewhere. It went down. That's compounding at work: losing 20% and gaining 20% doesn't bring you back to flat; it leaves you lower. Multiply that dynamic across dozens of swings, and "flat-ish volatility" can still translate into slow-motion damage.

A concrete mental model: you're buying the most expensive insurance plan in the store

Imagine an insurer that sells one-month catastrophe insurance. In calm times, the plan is priced with a premium because catastrophes are rare but devastating. You decide to buy the plan not for one month, but continuously—renewing it every day for years.

What happens? Most months, nothing catastrophic occurs. You pay premium after premium and get no payout. Occasionally, there's a storm and you get a payout—maybe a huge one. But if you're paying the richest premium in the store every month, your long-run experience is likely to be a net cost unless your timing is excellent.

That's the core of many long-vol ETP return profiles: most of the time, they are paying for protection. Sometimes they cash in. Over long holding periods, especially through calm regimes, the premium payments can dominate.

This is why "VIX products" are often described as tactical instruments. They can be very useful. They can also be financially radioactive when held with the wrong expectations.

Why ``the VIX stayed high'' doesn't guarantee your product did well

A classic retail confusion goes like this: "The VIX was elevated all year, so why did my long-volatility ETP still lose money?"

The answer is that the product wasn't holding "elevated VIX." It was holding a rolling set of futures that were priced above spot and bleeding through convergence, or it was rolling through a curve that stayed in contango even while spot VIX looked elevated. Spot can be sticky; the curve can still tax you. And the daily rebalancing can still drag you.

Also, volatility is not a single number; it's a surface. Spot VIX is a 30-day measure computed from SPX options. The futures curve is pricing where that 30-day measure will be *at future dates*. In an anxious but not panicked market, spot can remain elevated while the curve still slopes upward because the market expects normalization. That "normalization expectation" is exactly what hurts a long front-end futures roll strategy: it owns the part of the curve that is most eager to mean-revert.

So even if volatility feels high to you, the *carry* you are paying might still be high, because the market is charging you for *near-term* fear and expecting it to decay.

The widowmaker story: when the structure bites back

The term "widowmaker" gets thrown around in trading like a campfire ghost story. In volatility ETPs, it's not folklore. The contracts are designed with explicit rules that can turn a bad day into a terminal one.

The canonical example is the February 2018 blow-up in inverse VIX ETNs, where an inverse product experienced an acceleration event and was forced into redemption after an extreme move. Traders who treated it like a clever way to "short fear" discovered that the product wasn't a gentle short position they could manage; it was a note with a prospectus-defined kill switch. The trade didn't just lose a lot—it ended.

That event is worth mentioning for one reason: it reveals the true nature of these instruments. They are not just market exposures. They are market exposures plus *rules*. And rules matter most on the days you are least capable of reading fine print.

Even if you never touch inverse products, the lesson transfers: any volatility ETP can have hidden cliffs—rebalancing mechanics, liquidity constraints, creation/redemption disruptions, premiums/discounts, issuer risk (for ETNs). In calm markets, those cliffs look like harmless footnotes. In stress, footnotes become the main text.

The seduction: small losses that feel ``manageable"

Long-volatility ETPs rarely lose in a way that triggers your risk alarms early. They often grind down. A trader looks at a -8% drawdown and thinks, "That's annoying, but this is my hedge." Then it's -18%. Then it's -35%. At some point, the trader stops thinking of it as a hedge and starts thinking of it as a sunk cost they should recover.

That psychological pivot is where the product becomes dangerous. A hedge is something you can *pay for* and move on. A sunk cost is something you start to *justify*. And because the product can spike violently during the next volatility event, it's always dangling the most addictive possibility in trading: "I could get back to even if I just wait."

Meanwhile, the curve keeps billing you. Time becomes your enemy not because you're impatient, but because the product is literally designed to do work every day—rolling—whether or not your original reason for owning it still exists.

What a ``good" use of these products looks like

Long-volatility ETPs can be useful when you treat them like fire extinguishers:

- You buy them for a reason tied to timing (an event window, a risk regime shift, a portfolio vulnerability).
- You size them so the expected carry bleed is tolerable.
- You plan an exit when the reason expires, not when your P&L turns green.

The key is to stop confusing "liquid" with "safe." These products are liquid, yes. You can trade them easily. But their expected long-run return can be negative in the same way your expected long-run return from buying insurance every month is negative. That's not a bug. It's the business model.

A quick checklist that prevents most disasters

The goal is not to turn you into a lawyer; it's to keep you from being surprised by what you bought.

Before holding a volatility ETP beyond a very short tactical window, you want to know:

- Is it an ETN (issuer credit risk) or an ETF-style fund?
- Does it target short-term futures (front two months) or something longer-dated?
- Is it leveraged or inverse, and does it reset daily?
- How steep is the front-end contango right now (F2 minus F1)?

If you can't answer those in a minute, you're not trading volatility. You're trading a ticker and hoping the universe is kind.

Actionable takeaway: treat long-vol ETPs like rentals, not investments

Actionable takeaway: Write a "rental agreement" before you buy any long-volatility ETP. Put three items in writing: (1) the specific event or regime change you're hedging (not "the market feels risky"), (2) the maximum holding period you will allow if the spike doesn't happen, and (3) the maximum dollar amount you're willing to pay in carry bleed for the protection. If you can't set all three, the position isn't a hedge—it's an anxiety subscription with a ticker symbol.

4.5 Trading VIX Options

You buy a VIX call on a scary red day. The S&P 500 is sliding, your group chat is suddenly full of macro experts, and the VIX number on your screen is doing what it always does in movies: up and to the right. You feel clever for not panic-selling your portfolio. You've *hedged*. You've even picked a product that sounds perfectly matched to the problem: "volatility options."

Then the next day, the market wobbles again and the VIX prints even higher… and your VIX call barely moves. Or worse, it's down. The platform shows a loss and you stare at it like it's insulting your intelligence.

It's not insulting your intelligence. It's revealing the most important fact about VIX options: they are not options on the VIX you see on TV. They're options on a *particular VIX futures contract*, whether you realize it or not. If you trade them like they're tethered to spot VIX, you'll spend a lot of money learning the difference.

The underlying you think you bought vs. the underlying you actually bought

Most equity options traders have a clean mental map:

$$\text{AAPL call} \to \text{AAPL stock}$$

Even index options mostly behave like:

$$\text{SPX option} \to \text{SPX index}$$

VIX options break that intuitive wiring.

A VIX option's "underlying" is the VIX index level at settlement, but the VIX index level at settlement is tied to a very specific moment and methodology. In day-to-day trading, VIX options tend to behave as if they're connected to the *VIX futures market*, especially the contract month that matches the option's expiration cycle. Traders often summarize this with a blunt rule of thumb: *VIX options trade off the future, not spot.*

That rule of thumb is why your hedge can feel numb. Spot VIX can spike intraday on chaos, while the relevant VIX future moves less because the market expects the chaos to fade before the option's settlement date. Your option is pricing the market's expectation of volatility at a *future date*, not today's panic print.

If you only absorb one operational truth, make it this: when you buy a VIX option, you are making a bet on *where the market will price 30-day implied volatility at that option's settlement*, not on whether the next headline is ugly.

The curve strikes again: why ``back months don't flinch"

Earlier, the VIX futures curve was the reason carry exists. Here, it becomes the reason your option doesn't respond the way you expect.

When markets get hit, the front of the volatility complex usually reacts the most. That's where urgency lives. But VIX options are issued across maturities. A near-term VIX call (expiring soon) is more sensitive to an immediate volatility surge than a far-dated call. That part feels intuitive—short-dated options are "faster."

What's not intuitive is *how much* the sensitivity can collapse as you go out in time. Farther-dated VIX futures are often anchored by mean reversion. Even during violent selloffs, longer-dated futures can move less because the market believes volatility will cool over months. Your VIX option inherits that dampening. The option chain can look like a set of shock absorbers: the further out you go, the more the system absorbs today's bump.

This is why "buy a VIX call as a hedge" is not a single idea. It's a family of ideas. The maturity you choose determines whether you're buying a twitchy instrument that can pay quickly, or a slower instrument that may require a prolonged regime shift to respond meaningfully.

VIX options aren't SPY options with a scary label

VIX options are structurally different from the equity options most traders cut their teeth on. Two features matter immediately:

- They are *European-style*: you can't exercise early.
- They settle in a way that can surprise you if you hold into expiration.

European-style isn't a problem by itself. For many traders it's cleaner—fewer early exercise edge cases. The real issue is psychological: equity options traders are used to thinking "I can always manage it at the close." With VIX options, the *settlement value* is not a simple "whatever the VIX prints at 4pm." Settlement is tied to a special opening calculation on expiration morning. The option can stop trading, and then—after it's effectively too late for you to adjust—its final value is determined by a process that can diverge from the VIX number you watched the day before.

That's not a gotcha. It's the contract.

Settlement: the morning you don't control

The most dangerous moment in VIX options trading is not the day the market crashes. It's the morning your option settles and you assumed nothing special would happen.

VIX derivatives settle to a *Special Opening Quotation* (SOQ) derived from the opening process of the SPX options used in the VIX calculation. Translation: the final settlement value is built from an opening snapshot of the relevant options market, not from the VIX index level you saw bouncing around during regular hours. The settlement print can be higher or lower than what you expected based on the prior day's close, and the difference can be meaningful.

This has two practical consequences:

1. *"Pin risk" is replaced by "opening print risk."* With equity options, traders obsess about where the stock pins at the close on expiration day. With VIX options, your stress point is the morning settlement value.

2. *Managing the position at the close is not the same as managing settlement exposure.* You can be flat at Tuesday's close and still be exposed if you didn't understand when trading stops and how settlement is calculated.

If you've never traded something that settles off a special opening process, treat this as a different species. The contract's behavior at the end is not an exotic footnote. It's where many "I did everything right" stories go to die.

The hedge that bleeds: implied volatility can fall while you're ``right" about risk

VIX options have another unpleasant talent: they can lose value in exactly the environment where you feel you made the prudent choice.

Suppose you buy VIX calls as protection. The market sells off a bit, and you feel vindicated. But if the selloff is orderly—or if the market decides the bad news is contained—implied volatility can compress quickly. That compression hits VIX options hard because you're long implied volatility on implied volatility. It's leverage on a concept.

Also, because VIX options are linked to the futures curve dynamics, you can be correct that "today is scary" and still be wrong about "where this specific futures month will settle." In other words: your fear can be accurate, but mistimed relative to the contract's clock.

This is why VIX options are not just hedges—they're *timed hedges*. They want the fear to show up *before* the contract's horizon decays, and they want it to show up in the right part of the curve. If volatility spikes and then mean-reverts quickly, an option can feel like it watched the fire from behind glass.

A practical way to think about VIX option moneyness

Equity option traders have a comfortable concept of moneyness: compare the strike to the stock price. VIX options tempt you to do the same: compare the strike to spot VIX. That temptation is the source of many bad trades.

A more useful mental model is to compare the strike to the *relevant VIX future* level for that expiration month. If the future is at 19 and you buy a 25-strike call because spot VIX printed 23 intraday, you might be buying something much farther out-of-the-money than you think, priced off a calmer anchor. If you don't look at the future, you can't know what you actually bought.

Even if you never touch a futures contract directly, you should get into the habit of pulling up the futures curve before trading VIX options. It's not "extra research." It's checking what the option is functionally tethered to.

Greeks behave differently when the underlying mean-reverts

There's a reason VIX options feel alien if you try to trade them with the same reflexes you use on equities: the underlying process is different.

Stocks can trend. Volatility tends to mean-revert. That changes the character of delta and vega exposure. A VIX call can have a delta that behaves in a jumpy, regime-sensitive way—highly responsive during a panic, then quickly "deflating" as the market stabilizes. Vega (sensitivity to implied volatility) can also behave oddly because you're trading implied volatility on an index that is itself a measure of implied volatility. The nested nature matters.

In practice, that means you should be cautious about assuming a VIX call will behave like a long call on SPY. It often won't. The payoff can be more "event-like": dead, dead, dead, then suddenly alive. Or alive for an afternoon, then dead again when the curve relaxes.

That payoff shape is not a reason to avoid the product. It's a reason to trade it deliberately, with sizing and exits that respect the fact that mean reversion is always trying to erase your gains.

The expiration calendar: when ``one more day" becomes a different trade

Another subtle trap is assuming "monthly" means the same thing across markets. VIX options have their own listing conventions (including weekly listings under a different ticker symbol in many platforms). The calendar can be different from the equity options calendar you're used to. The last trading day and the settlement process can create a window where holding "just one more day" is not merely extending the same position—it's stepping into settlement mechanics you didn't intend to trade.

If your plan is to use VIX options as a hedge, the cleanest operational discipline is to avoid holding into expiration morning unless you *explicitly* want exposure to the settlement print. Many traders don't. They just drift into it.

What VIX options are actually good for

VIX options shine when you match them to the problems they're built for:

- *Event hedging:* known windows where volatility could reprice sharply.
- *Tail insurance:* defined-risk convexity with no margin calls (premium paid up front).
- *Volatility views:* expressing a belief that implied volatility for a particular horizon is mispriced.

They are less suited to the lazy version of hedging: "I want something that always goes up when the market goes down." That product doesn't exist in clean form. What exists are instruments with clocks, curves, and settlement rules—tools that can be incredibly sharp when used properly, and surprisingly dull when treated as symbolic protection.

Common mistake: buying too far out-of-the-money, too far out in time

A VIX call that's far out-of-the-money is cheap for a reason. It needs a truly violent volatility event to pay. Combine that with a back-month expiration that's dampened by mean reversion, and you can buy a product that feels like insurance but behaves like a lottery ticket.

That doesn't mean you should never buy far out-of-the-money VIX calls. It means you should label them honestly in your own mind. If you're buying a lottery ticket, fine—size it like one. If you're buying a hedge, build it so it has a reasonable chance of helping in the scenario you're actually afraid of.

A hedge that only pays in an apocalypse is emotionally comforting and financially useless.

Actionable takeaway: a pre-trade checklist that prevents the classic VIX option disappointment

Actionable takeaway: Before buying any VIX option, do three quick checks and write the answers in your notes: (1) which expiration are you trading and what is the matching VIX futures month price right now, (2) are you comfortable with the contract's expiration/settlement mechanics (special opening settlement) or will you exit before that window, and (3) what exact market scenario makes your option pay—"spot VIX spikes today" is not specific enough; you want "the relevant futures month reprices higher by X points within Y days." If you can't define (3), you're not hedging—you're hoping the ticker behaves like a movie.

4.6 Behind the Scenes: Variance Replication

A funny thing happens when people first learn the VIX is "just math from options." They immediately imagine a single magical option—the one contract that contains all the market's fear. They picture a lever labeled *VIX* that someone pulls when the world feels shaky.

Reality is stranger and more interesting: the VIX is built from a *crowd*, not a hero. It's computed from a wide strip of S&P 500 index options across many strikes. The VIX is what you get when you take a whole options chain—especially the out-of-the-money parts that most people ignore in calm markets—and blend it into one number that represents the market's priced expectation of variance over the next month.

That word *variance* is the secret door. If you walk through it, the VIX stops being a mysterious fear gauge and starts looking like something almost physical: an index that behaves the way it does because of how risk is distributed across strikes, and because tails—the ugly, far-out scenarios—matter disproportionately.

Volatility vs. variance: why the square matters

Volatility is the number traders speak in: "VIX is 18," "implied vol is 22." Variance is what the math likes: volatility squared.

That sounds like pedantry until you notice how payoffs behave. A portfolio can have daily wiggles that cancel out directionally, yet still produce a lot of *movement*. Variance is a clean way to measure movement because it adds up naturally over time. If you double the time horizon, expected variance roughly doubles (all else equal). Expected volatility does not; volatility scales with the square root of time. That's why traders do $\sigma\sqrt{T}$ in their heads, and why the VIX is quoted as an annualized number even though it's trying to describe roughly 30 days.

When you hear "VIX is 20," that's volatility language. Under the hood, the machinery is estimating expected *variance* over the horizon, then taking the square root and annualizing it to express the result in volatility units. The square-root step is why the VIX feels intuitive ("20% annualized vol") even though the replication logic underneath is a variance story.

This variance framing is also why VIX-linked products can be slippery: you're not trading a simple price. You're trading a market-implied second moment—a statistical property—extracted from a complex surface of option quotes.

The intuition: building ``movement exposure" without forecasting direction

Imagine you want a bet that pays when the market is choppy, regardless of whether it chops up or down. That's the core idea of a variance swap: a contract that pays based on realized variance over a period. It's like saying, "I don't care where the S&P 500 ends up; I care how violently it gets there."

Now here's the elegant part: you can approximate that kind of "movement exposure" using a *strip of options* across strikes. Not one option. A lot of them.

Why does that work? Because options are sensitive to the distribution of future prices. A deep out-of-the-money put is a tiny bet on a disaster. A slightly out-of-the-money put is a bet on a smaller disaster. Out-of-the-money calls capture upside tails. If you combine enough of these options with the right weights, you can create a position that behaves like "pay me for variance." It's model-free in spirit: you're not assuming the world follows a perfect bell curve with one volatility parameter. You're letting the market's own option prices describe the distribution, then extracting the variance implied by that distribution.

The VIX is essentially a standardized, exchange-published version of that idea. It uses a broad set of SPX options, weighted in a particular way, to estimate the market's expected variance for a constant-maturity window around 30 days.

Why a strip, not a single at-the-money option?

Newer traders often assume at-the-money options are "the most important" because they're closest to the current price and trade the most. At-the-money options are important, but they're also incomplete. They're like a microphone pointed at the center of the crowd. You hear the loudest voices, not the whispers that become screams in a crisis.

Variance is influenced by the whole distribution, including tail outcomes. Deep out-of-the-money options—especially puts—encode the market's pricing of rare but brutal moves. In quiet times, those options can be cheap and ignored. In stressed times, they can reprice dramatically as investors scramble for crash insurance and dealers demand compensation to warehouse that risk.

Because the VIX draws from a wide range of strikes, it "listens" to that repricing. That's why VIX spikes can feel disproportionate to what just happened in the index. The market may not have moved *that* much yet, but the price of tail insurance may have exploded. The VIX is sensitive to the cost of insuring the tail, and it will reflect that sensitivity.

This is one reason the VIX can act like it has a hair trigger: it's not measuring today's return. It's measuring the price of the distribution of returns, and distributions can change shape fast when liquidity thins and hedging demand surges.

A memorable rule: tails punch above their weight

In the replication logic behind variance, low-probability outcomes can have high impact. A 1% chance of a huge move contributes more to variance than a 1% chance of a tiny move. That's obvious once you say it out loud, but it has a powerful consequence: the market can become dramatically more expensive to insure against rare crashes even when the "most likely" outcome hasn't changed much.

That's why the VIX is so sensitive to the left tail of the equity market. When investors rush to buy puts, they are not just expressing "fear." They're changing the market price of tail outcomes, and tail outcomes have a leveraged effect on expected variance.

If you've ever watched the VIX surge on a day where the S&P 500 is down "only" 1.5% and thought, "This seems excessive," what you're really seeing is the market repricing the probability and/or magnitude of bigger moves *ahead*. The VIX is less a reaction to the punch you just took and more a price tag on the punches you're now worried might be coming.

Microstructure: when quotes, spreads, and liquidity become the story

There's an important real-world caveat that keeps the variance replication story honest: the VIX is computed from option *quotes* (bid/ask), and quotes can behave differently than trades—especially in stress.

In calm markets, the difference is minor. Tight spreads mean quotes are informative and stable. In chaos, spreads widen.

Market makers step back or reprice aggressively. The options chain can become a landscape of gaps and jumpy quotes. Because the VIX calculation consumes those quotes, the index can jump in response to microstructure—widening spreads and shifting dealer behavior—not only in response to investors trading huge volumes.

This matters because it explains some of the VIX's most confusing prints: sharp spikes that appear to come "out of nowhere," or intraday surges that partially retrace once liquidity improves. The index is faithfully reflecting the cost to transact in the options that define it. When the cost of transacting becomes temporarily distorted by stress, the VIX reflects that distortion. The index is not lying; it's reporting a market that is briefly hard to price.

For a trader, this is both warning and opportunity. Warning, because your mark-to-market may be driven by liquidity conditions more than fundamental information. Opportunity, because if you understand how liquidity and dealer quoting can exaggerate implied variance, you can be more selective about when you chase a spike versus when you fade it.

The constant-maturity trick: always pointing at ``about 30 days"

The VIX doesn't lock onto a single expiration and stubbornly stick with it until expiry. It targets roughly 30 days, continuously. That means as time passes, the option maturities used in the calculation change, and the index "rolls" its measurement window forward.

This constant-maturity design is why the VIX can be a cleaner indicator of near-term implied volatility than just looking at "the next SPX option expiration." It also means the index is always drawing from a zone of the options chain where liquidity and event risk can vary. Around major macro events, the market may stuff extra implied variance into the 30-day window, and the VIX will capture that. After the event passes, the same window can collapse. The VIX is always measuring the cost of insuring the next month, not the mood of the last month.

Why the VIX can feel ``jumpy" compared to realized volatility

Realized volatility is a historian. It records what happened—often using returns sampled daily. The VIX is a novelist writing a draft of the next month using live option prices. Novelists can change their minds mid-paragraph.

Because options embed expectations about future moves, the VIX can jump before realized volatility has time to catch up. It can also overshoot and then mean-revert quickly. That behavior is not just psychology; it's the market constantly repricing the distribution of outcomes as new information arrives and as positioning changes.

Also, realized volatility has a smoothing effect: it accumulates over days. The VIX can react in seconds. When the market suddenly prices a fatter tail, implied variance can reprice instantly. Realized variance can't. It needs actual movement over time. This timing mismatch is exactly why volatility trading is hard and fascinating: the market is pricing the storm while the weather is still forming.

A practical trading implication: VIX is a tail-sensitive index, not a ``normal move'' meter

Many traders use the VIX as if it were simply the market's expectation of "typical" movement. It's partly that—but its construction makes it particularly sensitive to the pricing of crash protection. That sensitivity is one reason the VIX tends to have an asymmetric relationship with equities: it explodes during drawdowns and drifts lower during rallies.

So if you're trading VIX-linked instruments, it's useful to think in distributions rather than point forecasts. Ask yourself:

- Is the market repricing the *tail*, or just the center?
- Are put spreads widening because investors are urgently hedging, or because liquidity is thin?
- Is implied variance rising because the world got riskier, or because the market got harder to price?

Those questions are not philosophical. They help you avoid paying top dollar for protection *after* the repricing has already occurred, and they help you understand why some spikes fade quickly once liquidity returns.

Putting it all together: why VIX ``feels emotional'' even though it's mechanical

The VIX is computed. It's mechanical. Yet it behaves in a way that maps eerily well to human emotion—spiking sharply, calming slowly, flaring around uncertainty. That's because option prices are where emotion and money meet. Fear isn't an input to the formula, but it's an input to the bids and offers that the formula consumes.

And because the formula listens to tails, it listens hardest to the part of the market that gets loudest when humans get stressed. When people say "VIX is fear," they're noticing a correlation. When you understand variance replication, you understand the causality: the market reprices tail insurance, tail insurance has outsized weight in expected variance, and the VIX reports that repricing in real time.

That's not a nickname. That's a mechanism.

Actionable takeaway: stop watching one number---watch the tail

Actionable takeaway: The next time the VIX spikes, don't just stare at the index level. Pull up an SPX options chain (or a volatility surface view) and look specifically at out-of-the-money puts: are their implied vols jumping more than at-the-money vols, and are bid/ask spreads widening? If the tail is repricing and spreads are blowing out, the VIX move is likely being driven by tail insurance becoming scarce (and potentially by microstructure). That tells you two things immediately: (1) VIX-linked products tied to the front end can move violently, and (2) chasing the spike can be expensive if liquidity normalizes. You don't need to compute the VIX formula to trade it better—you just need to remember that the tail is where the index gets its punch.

Chapter 5

Reading the Room: Skew, Smile, and Sentiment

*If the VIX is the market's thermometer, the volatility surface is its MRI scan. A single volatility number only tells you the average temperature; it doesn't tell you *where* the inflammation is hiding. In this chapter, we stop treating volatility as a flat number and start seeing it as a terrain. We'll explore why the market grimaces to the downside, why 'crash insurance' is priced like a luxury good, and how sophisticated traders use the shape of the volatility curve-not just its direction-to find the best deals on the board.*

5.1 The Smirk: Why the Market Grimaces to the Downside

Imagine you're shopping for two kinds of insurance on the same house.

One policy pays if your house catches fire *today*. The other pays if a meteor lands on it *this year*. Both are "unlikely," both are ugly, both would be catastrophic. Now watch what happens when you try to buy them in the real world: the price won't be set by some tidy spreadsheet that says "unlikely equals cheap." It'll be set by the oldest force in markets: who *needs* the protection, and how many people can stomach writing the check when the worst day arrives.

That's the right mental picture for equity options. The market isn't evenly afraid of up moves and down moves. It doesn't price a neutral, symmetrical world where surprises are equally likely to be delightful or horrific. Instead, it prices something closer to lived experience: stocks can drift upward for months and then fall down a staircase while everyone reaches for the same exit door at once. The result is the defining facial expression of equity implied volatility: not a "smile," but a *smirk*.

A quick translation, in plain language:

- *Implied volatility* (*IV*) is the volatility level that makes an option's market price "make sense" inside an option pricing model. It's not a forecast you can hold in your hand; it's the market's *price tag* for uncertainty.
- A *volatility smile* is what you get when options far from the current price (far out-of-the-money calls and puts) trade with higher IV than at-the-money options, often in a roughly symmetric U-shape.
- *Equity index markets usually don't smile. They smirk.* The left side (out-of-the-money puts) is lifted higher than the right side (out-of-the-money calls). Downside IV is rich; upside IV is comparatively cheap.

If you've ever looked at an option chain and wondered why "protect me from a crash" costs so much more than "let me chase a melt-up," the smirk is the answer in one picture.

The clean, theoretical world begins with a simple idea: returns are nicely behaved, and uncertainty is evenly distributed. In that world, deep out-of-the-money options on either side are "rare events," so both wings of the distribution are priced similarly. That's where the textbook smile comes from.

The real equity world is different for reasons that are not philosophical. They're structural. They're institutional. They're mechanical. And once you see them, the smirk stops looking like a quirky chart artifact and starts reading like a crowd's body language.

Why puts are expensive (and why the market keeps paying anyway)

Out-of-the-money (OTM) puts are expensive because lots of large, serious players *must* own them, or at least must behave like they might need them on short notice.

Consider who shows up as a natural buyer of downside protection:

- Pension funds and endowments with mandates that effectively say: "We can't be down 30% right when our liabilities come due."
- Risk-parity and volatility-targeting strategies that reduce equity exposure when volatility rises, which can *accelerate* selling during stress.
- Asset managers judged on drawdowns, not on how cleverly they explained them.
- Corporate buyback desks and structured product issuers who, in certain regimes, interact with option mar-

kets in ways that can intensify demand for downside convexity.

Now consider who is a natural seller of crash insurance. On paper, it's "anyone who wants premium income." In practice, it's whoever can survive the day the premium is revealed as not income but a down payment on pain.

Selling OTM puts is like being the flood insurer in a town that's been dry for years. You can collect checks for a long time. People will call you smart, disciplined, even "data-driven." Then the river jumps the bank, and the math stops being polite. In that moment, the seller is asked to be a liquidity provider, a shock absorber, and a therapist for the market's fear—all at once. There simply aren't enough balance sheets willing to do that cheaply.

So the price of OTM puts stays high. Not because every day is a disaster, but because the market remembers that the bad day arrives *without asking permission*.

Right there is the intuition behind the smirk: the left tail isn't just "less likely." It's *more crowded*. When fear shows up, it doesn't show up as a polite increase in uncertainty. It shows up as a stampede into the same strikes, the same expirations, the same trades.

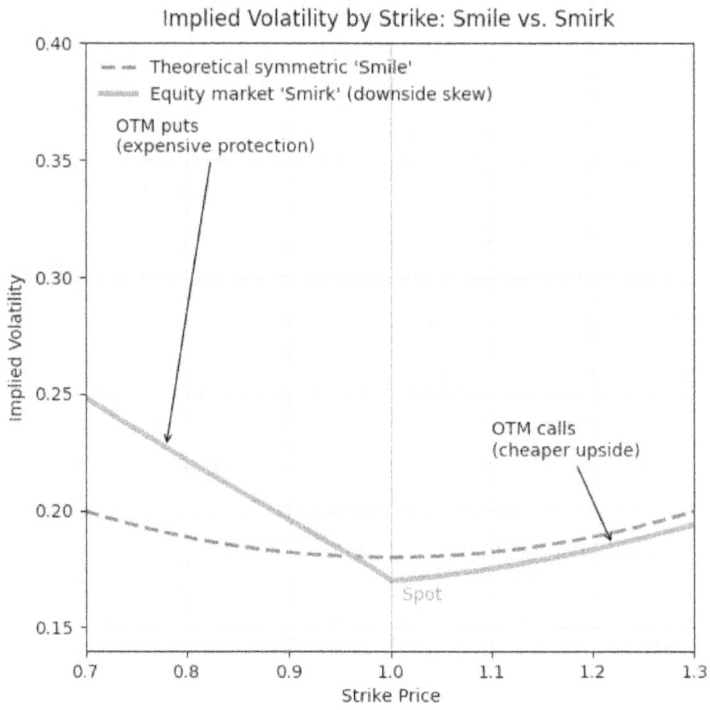

Skew is not a prediction; it's a price of urgency

A common trap is to treat skew like a weather forecast: "Puts are expensive, so a crash must be coming." Sometimes that's true. Often it isn't. Skew is better understood as the market charging more for *urgency* on the downside than on the upside.

Think about the timing of pain.

When the market grinds higher, it usually does so with time to react. Committees meet. Stops get adjusted. Risk gets trimmed with a day's notice. The upside tends to be an *appointment*.

When the market breaks lower, it tends to do so with the grace of a dropped glass: one second you're holding it, the next second you're cleaning up the shards. The downside is an *ambush*.

Options are valuable when you don't have time. That's why OTM puts carry a premium that feels almost rude. They're selling you something scarce: the ability to make a decision calmly on a day when calm is expensive.

The ``leverage effect": why falling prices create rising volatility

The smirk also has a mechanical backbone: in equities, volatility tends to rise when prices fall. That negative relationship is often called *spot-vol correlation* (spot price versus volatility), and one story for it is the *leverage effect*.

Here's the intuition without the algebra. If a company's equity price drops while its debt stays the same, the equity becomes a thinner cushion on top of the debt. The company is effectively more "levered," and the equity becomes riskier. Riskier equity tends to mean higher volatility.

Even if you don't love that corporate-finance story, the market-behavior version is hard to argue with: selloffs bring forced selling, margin pressure, and risk limits that bite at exactly the wrong time. Dealers hedge. Funds de-gross. "Vol control" cuts exposure. The market becomes a machine that turns down moves into more volatility demand.

Now connect that to the smirk: if everyone knows volatility tends to rise in a drop, then puts aren't just protection against price declines. They're protection against the *volatility spike* that makes the decline feel like it has teeth. That extra feature—the tendency for fear to amplify itself—makes puts structurally more valuable.

Skew is a supply-and-demand chart wearing a math costume

It's tempting to talk about skew as a sophisticated statistical object: implied distributions, negative skewness, fat tails. Those are real concepts. But you can understand 80% of what you need by watching who is leaning on which side of the boat.

Downside puts are where institutional fear gets expressed. Not because institutions are emotional, but because they have constraints. A CIO can survive being underinvested in a rally. A CIO may not survive a catastrophic drawdown that forces spending cuts, contribution hikes, or redemptions. So they pay for convexity.

Upside calls are more often a discretionary desire. People want them. But fewer people *need* them. Upside exposure can be expressed by simply owning the underlying. Downside protection cannot.

That asymmetry— "need" versus "want" — is the smirk's permanent generator.

There's another subtle point: the put wing can become expensive even when nobody is actively "bearish." A market can be optimistic and still pay up for puts because optimism doesn't cancel the career risk of a crash. In fact, the more complacent the environment feels, the more treacherous the sentence becomes: "We don't need hedges right now." Many professionals have learned that lesson the hard way, so demand for puts never truly disappears; it just hibernates.

How to *read* the smirk: richness, steepness, and what traders actually compare

When people say "skew," they might mean a few related things:

- *Put IV versus call IV at comparable distance from spot.* Example: compare a 5% OTM put IV to a 5% OTM call IV in the same expiration.
- *IV slope across strikes.* How quickly IV rises as strikes move lower.
- *Delta-based comparisons.* Traders often compare a 25-delta put IV to a 25-delta call IV because "25-delta" is a standardized way to talk about moneyness across different vol regimes.

The practical read is this: *steeper left skew means the market is charging a bigger premium for crash convexity relative to everything else*. It's not just "vol is high." It's "vol is *unevenly* high in the places that pay when things go wrong."

Two details make this more useful than it first appears.

First, skew can change even when at-the-money volatility doesn't. You can have a calm tape, a stable VIX-ish regime, and still see put wing IV inflate. That's the market quietly restocking disaster insurance without necessarily bidding up the whole front end.

Second, skew can flatten during violent selloffs. That sounds backwards until you visualize what happens: when panic hits, at-the-money volatility can explode higher so quickly that the entire curve lifts, sometimes compressing the relative difference between wings. The smirk doesn't always get "steeper" in a simple way; sometimes fear grabs the whole face, not just the left corner of the mouth.

So the right habit is comparative: *don't stare at one number; compare shapes through time*. Ask:

- Is the whole curve lifting (vol level), or is the left wing getting especially rich (skew)?
- Is the market paying for *immediacy* (near-dated skew) or *structural tail risk* (longer-dated skew)?

A non-obvious reason the smirk persists: the ``volatility risk premium" is not evenly distributed

There's a simple, almost uncomfortable truth that sits beneath most option markets: on average, option buyers tend to overpay relative to subsequent realized volatility. That tendency is often discussed as the *volatility risk premium* (the compensation sellers demand for warehousing volatility risk).

But in equity indices, that premium isn't spread evenly across strikes. The left tail is where sellers demand the most compensation. Why? Because left-tail losses are exactly the sort that arrive with bad liquidity, broken correlations, and risk managers calling at the worst time. The market isn't just pricing variance. It's pricing the experience of trying to manage risk when the building is on fire.

If you want one sentence that captures why the smirk is persistent, it's this:

The market charges more for protection that pays when everything else stops working.

OTM puts tend to do exactly that. In a crash, correlations go to one, diversification evaporates, and "safe" becomes a crowded trade. Downside convexity is one of the few assets that can still feel liquid and effective when the rest of the toolkit is failing. That's why it carries a premium that doesn't go away simply because the last few months were calm.

What skew is telling you about sentiment (and what it isn't)

It's seductive to treat skew like a lie detector for fear. Sometimes it is. But it's more reliable as a detector of *constraints*.

When downside skew is very steep, it can mean:

- A wave of hedging demand (protection buying).
- A shortage of natural sellers (balance sheets pulling back).
- A market that expects spot-vol correlation to bite hard in a drop.
- A "gap risk" environment where traders fear discontinuous moves (overnight, macro shocks, policy surprises).

What it does *not* necessarily mean:

- A crash is imminent.
- You should automatically sell puts because "they're overpriced."

- Upside calls are "cheap" in an absolute sense (they can still be expensive; they're just cheaper relative to puts).

The best way to keep your head straight is to remember that skew is a relative price. It's the market shouting: "I am more stressed about the left tail than the right tail." That's a statement about risk appetite and positioning as much as it is about expected outcomes.

How traders use the smirk without getting hypnotized by it

The smirk becomes useful when you treat it like a menu, not a prophecy.

If downside protection is structurally expensive, then buying it constantly is like paying for valet parking every day. It's convenient. It's sometimes brilliant. It's also a persistent drag unless you time it well or you truly need the service.

So traders and risk managers often do one of three smarter things than "always buy puts":

- *Define the disaster you can't survive, then hedge that, not everything.* If a 10% drawdown is uncomfortable but survivable, why pay for protection that begins paying at 2% down every month? Skew punishes vague hedging.
- *Use structure to reduce premium bleed.* Instead of buying a naked put, consider spreads (buy one put, sell a lower strike put) so you're paying for the part of the left tail you actually care about.
- *Trade relative value in shape.* Sometimes the market overpays for the *far* left tail relative to nearer strikes, or vice

versa. That's where skew becomes a tradable object, not just a scary chart.

You don't need to be a derivatives savant to apply this. You just need to stop thinking of options as "bets" and start thinking of them as *insurance contracts with wildly different deductibles and premiums*. The smirk is the insurance market's pricing table.

Actionable takeaway

The next time you pull up an option chain, do this one-minute read before you place any trade:

1. Pick a single expiration (keep maturity constant so you're reading skew, not time).
2. Compare an OTM put and an equidistant OTM call (or, better, 25-delta put IV versus 25-delta call IV).
3. Ask a practical question, not an academic one: *"How much extra am I being charged for downside urgency?"*

If the left wing is dramatically richer, treat downside hedges like catastrophe insurance: buy them only with a clear definition of what you're protecting, and consider structures that cap the premium bleed. If the smirk looks oddly muted, don't celebrate; get curious. Sometimes the market is calm. Sometimes it's complacent. The difference shows up first in the places people *stop* paying for protection.

5.2 The Surface: A 3D Map of Fear

A one-day option chain is a persuasive liar.

It shows you a clean row of strikes, tidy implied vols, and the comforting illusion that "the market" has a single opinion about risk. But the market doesn't think in one dimension. It doesn't even think in two. It thinks in *when*.

A 10% drop tomorrow and a 10% drop sometime in the next two years are both unpleasant, but they don't feel the same in your stomach or in your margin account. One is a car crash. The other is arthritis. Options price both, and they price them differently, because time changes everything: the probability of a move, the path it might take, and the number of headlines that could show up and shove the market around.

So the skew "smirk" is only a slice of a larger object. Add expiration, and the picture turns from a line into terrain: peaks, ridges, and cliffs. Traders call it the *volatility surface*.

From a smirk to a landscape

Start with the simplest view: pick one expiration date and plot implied volatility against strike price. That's the smirk you've already met—a left wing that's lifted like a raised eyebrow, because downside insurance is scarce and urgently desired.

Now do something that sounds almost boring but changes your whole perspective: plot that same curve again for a different expiration. Then again for a longer one.

Suddenly you're staring at a family of smirks stacked into the future. Some are steep and nervous. Some are flatter and sleepy. Some tilt in odd ways around known events. And the interesting part isn't any one curve—it's the relationship between them.

Time-to-expiration (often called *tenor*) is the second axis that turns "skew" into a surface. The third axis is implied volatility itself. The result is a 3D map of how the market prices uncertainty across both *where* (strike) and *when* (maturity).

If you manage money, the surface is closer to the truth than any single headline volatility number. If you trade options, it's where edge lives, because mispricings tend to hide in the joints: between near-term and long-term, between at-the-money and wings, between "everyone is staring here" and "nobody bothered to compare that."

The surface has weather and climate

Here's a practical way to keep the intuition vivid:

- Near-dated implied volatility is *weather*. It can change in an hour. It flinches at earnings, CPI prints, surprise press conferences, and a bad tape at 3:47pm when liquidity thins out.
- Long-dated implied volatility is *climate*. It moves too, but it tends to move with slower forces: recession risk, regime shifts, credit stress, or a market that quietly decides the next few years will be messier than the last few.

The market can be calm about next month and uneasy about next year. It can also be terrified about next week and relatively chill beyond that, because panic is often acute even when the long-term story hasn't changed.

That difference matters because it tells you what kind of fear you're dealing with. Acute fear is about *timing*. Structural fear is about *distribution*.

And the surface shows you both at once.

Term structure: the curve of time

When people say "term structure" in volatility, they're usually talking about how implied volatility changes with expiration, holding moneyness roughly constant (for example, comparing at-the-money IV across maturities).

Two shapes show up so often they deserve to feel familiar:

- *Contango*: longer-dated implied vol is higher than near-dated implied vol. This is common when markets are relatively calm. The front end is sleepy; the back end carries a "just in case" premium.
- *Backwardation*: near-dated implied vol is higher than longer-dated implied vol. This tends to happen during stress. The market isn't worried in the abstract; it's worried *right now*.

Those words get used heavily in VIX futures, but the idea is bigger than the VIX complex. It's a general statement about the pricing of uncertainty through time. The surface lets you see whether fear is an urgent, near-term demand for protection or a slower, persistent repricing of risk.

A subtle but important point: contango isn't "bullish" and backwardation isn't "bearish." They're *organizational charts for fear*. Contango says, "today is fine, but the world is never perfectly fine." Backwardation says, "today is *not* fine."

Skew plus term structure: why the surface is more honest than a single number

A single implied vol number—even a famous one—is a compression algorithm. It throws away detail. That's not a flaw; it's a feature. You want compression sometimes. But if you only look at the compressed summary, you can miss the story.

Here's the kind of story the surface can tell that a single number cannot:

- The market may not be bidding up at-the-money vol much, but it may be quietly paying up for *downside tails* in longer-dated tenors (a "slow dread" regime).
- The market may be panicking in the front month but not changing the back end much (a "headline shock" regime).
- The at-the-money vol may lift everywhere, but skew may flatten (fear is broad, not targeted).

If skew is *where* fear is concentrated, term structure is *when* fear is concentrated. The surface is the combined answer.

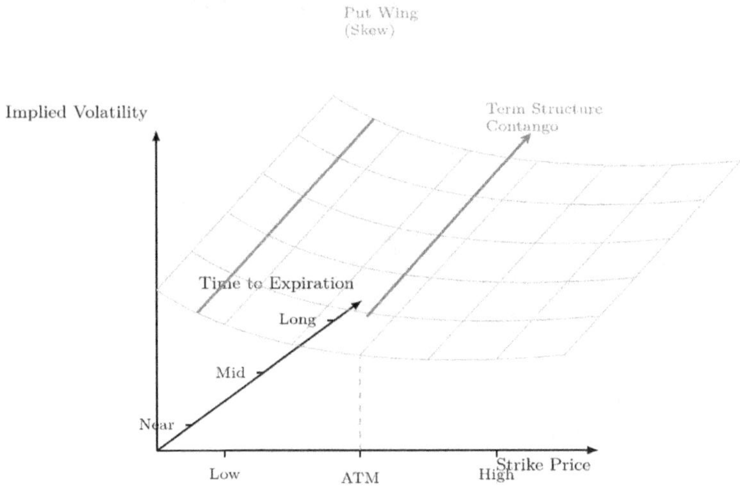

Why the surface is where trades become ``about pricing,'' not ``about prediction''

A lot of people come to options looking for a fancy way to bet on direction. That's the least interesting use of the instrument. The interesting use is to bet on *how everyone else is paying for uncertainty.*

The surface is where that becomes visible.

Suppose you see steep downside skew in near-term options, but the longer-dated skew is comparatively mild. That can happen when the market is bracing for an event risk in the next few weeks—something like a binary macro catalyst or a cluster of earnings—but not repricing long-term crash risk much. The fear is sharp, not deep.

Flip it: a relatively calm front end with persistent, expensive long-dated put wings. That's the market expressing a different mood: "tomorrow is probably fine, but the distribution over the next year is uglier than it used to be." That kind of slow repricing often doesn't get the adrenaline of a VIX spike, but it can matter more for portfolio hedging decisions because it changes the carrying cost of long-horizon protection.

The surface also helps you avoid a classic beginner mistake: comparing apples to meteorites. People will say, "the 10% OTM put is expensive," while looking at a 7-day put, then compare it to a 60-day call and conclude the market is "bearish." That's not market psychology; that's mismatched tenors. The surface forces you to compare like with like.

Where the surface gets weird (and why weirdness is valuable)

In quiet markets, surfaces often look pleasantly regular: higher IV on the put wing, a gentle term structure, no dramatic wrinkles. In real trading life, surfaces develop *scars*. Those scars are where opportunity and danger show up together.

Some common sources of wrinkles:

- *Event humps*: a single maturity trades rich because it spans a known event (earnings, a central bank meeting, an index rebalance that traders care about). The surface bulges at one point in time.
- *Wing kinks*: a particular delta region trades oddly rich because flows are concentrated there (protective put demand, structured product hedging, systematic overwriting). The surface develops a ridge.
- *Calendar dislocations*: the front is cheap relative to the second month, or vice versa, beyond what you'd expect

from simple mean reversion. This is the options market saying "timing" matters more than the headline level.

The surface is valuable precisely because it can look wrong. A perfectly smooth surface is a world where nobody has an urgent need, no constraints, and no deadlines. That world does not trade.

A trader's way to think about ``cheap" and ``rich" on the surface

Calling an option "cheap" or "expensive" in isolation is like calling a hotel "overpriced" without mentioning whether it's New Year's Eve.

Options are *relative goods*. Cheap relative to what?

On the surface, "rich" usually means:

- Rich relative to nearby strikes (skew is steep; the wing is being bid).
- Rich relative to nearby maturities (one point on the term structure is inflated).
- Rich relative to realized volatility expectations (you believe the market is paying too much for what will actually happen).

The first two are surface-native comparisons. They don't require you to forecast realized volatility with heroic precision. You're trading the market's internal inconsistencies: shape versus shape, month versus month, wing versus belly.

This is where many practical volatility strategies live:

- *Calendars and diagonals* express a view about time: near-term vol relative to long-term vol, often with minimal directional exposure if structured carefully.
- *Vertical spreads* express a view about skew: the relative pricing of different strikes in the same maturity.
- *Structures that "buy what's cheap and sell what's rich"* are, at their core, surface trades.

You don't need to memorize Greek letters to start thinking this way. You just need the habit of asking: "Am I paying for weather when I only need climate, or paying for climate when the storm is actually tomorrow?"

How to read the surface like a risk manager (not just a trader)

There's a quieter use of the surface that matters even if you never place a single options trade: it tells you what kind of risk the market is most sensitive to *right now*.

- If the front end is bid and the back end is stable, the market is jumpy—a headline-driven environment where liquidity and gamma matter.
- If the back end lifts and skew steepens across maturities, the market is repricing regime risk—the possibility that correlations, growth, or policy uncertainty are changing for longer than a week.
- If skew is extreme but at-the-money vol is only modest, the market is paying for tails without necessarily expecting daily chop—a world where the average day is calm, but the bad day is feared.

That's risk intelligence. It can influence how you size positions, where you set stop levels, how aggressively you rebalance, and whether your "diversifiers" will actually diversify when it matters.

The surface won't tell you what will happen. It will tell you what would hurt.

Actionable takeaway

Do a five-minute "surface check" before you trade, hedge, or even rebalance a portfolio:

1. Pick *three expirations*: near (around 1–2 weeks), mid (around 1–2 months), and long (around 6–12 months).
2. For each expiration, record implied vol for *three points*: an at-the-money option, an OTM put (roughly 25-delta), and an OTM call (roughly 25-delta).
3. Answer two questions in plain English:
    - Is fear concentrated in *time* (front month bid vs back month), or in *tail* (put wing bid vs everything)?
    - Is the market charging you for *weather* (near-term chaos) or *climate* (longer-term uncertainty)?

Write the answers down. After a few weeks, patterns jump off the page. You'll start to recognize when you're paying a premium for immediacy, when the market is quietly repricing long-horizon risk, and when a "simple" options quote is actually a point on a much larger map.

5.3 Panic Dynamics: The Escalator and the Elevator

At 10:13 a.m. on an ordinary trading day, a portfolio can look like a well-built bridge: diversified, balanced, engineered to survive wind and weather. By 10:43 a.m. on a bad day, that same portfolio can feel like a folding chair.

Nothing "new" happened to the long-term value of most companies in that half hour. What changed was the market's demand for *certainty*. And in options, that demand doesn't express itself politely. It expresses itself as a sudden, violent repricing of *downside convexity*—the property that makes puts accelerate in value when prices fall and volatility rises.

That's the escalator-and-elevator dynamic: stocks tend to climb on an escalator (grinding, stepwise, with pauses), but they can drop in an elevator (fast, vertical, with your stomach lagging behind the floor). Volatility does the opposite. It dozes on the way up and wakes up screaming on the way down.

The secret handshake between spot and vol

In equity indices, the relationship between price and implied volatility is often negative: when the index drops, implied volatility tends to rise. Traders call it *spot-vol correlation*. You don't need a correlation coefficient to feel it. You've felt it anytime you watched a market dip and saw puts suddenly "wake up" even if the expiration date didn't change.

Three forces make this relationship feel almost baked into the wiring:

- *Mechanical risk*: falling equity prices can increase financial leverage (the equity cushion gets thinner), which can make the equity behave more wildly.
- *Flow risk*: selloffs trigger hedging demand, risk limits, and forced deleveraging. Those flows are pro-cyclical: they tend to intensify the move.
- *Liquidity risk*: when markets fall hard, bid-ask spreads widen and depth vanishes. Options become more expensive partly because it's riskier to make markets in them.

In calm conditions, you can pretend these are separate. In panic, they braid together. Price falls increase volatility; higher volatility increases the desire (and sometimes the need) to hedge; hedging flows can push prices further; liquidity thins; and the cost of insurance jumps again.

This is why the market's facial expression changes in a crash. It's not just that implied volatility rises. It rises in a lopsided, angry way.

Skew doesn't just lift; it warps

In quiet markets, downside skew is a kind of steady background hum: OTM puts are richer than OTM calls, but the curve is readable, continuous, almost civilized.

In a crash, the surface doesn't simply shift upward like a tide. It *warps*.

Here's what that warping often looks like in plain language:

- At-the-money implied volatility jumps because the market expects bigger daily swings *immediately*.
- The put wing gets bid even harder because the market starts caring about *gaps*, not just moves. A 2% down day is one thing. A 6% overnight gap is another.
- The steepness of the left tail often increases because the demand for protection becomes less *portfolio hygiene* and more *oxygen mask*.

That last point is where many traders get surprised. They expect volatility to rise during a selloff, but they underestimate how much of that rise can be concentrated in strikes that looked irrelevant a week earlier.

A deep OTM put is easy to dismiss when it's cheap. In a crash, it stops being a lottery ticket and starts being a lifeboat. The same instrument, two different identities, depending on whether the market thinks it's pricing *variance* (normal choppiness) or *tail* (discontinuity).

Why the left tail is where panic shows its teeth

Panic is not symmetrical. Humans don't panic about "missing out" the way they panic about "ruin." Institutions are even more one-sided, because they're governed by constraints: drawdown limits, VaR limits, margin rules, client redemption risk, and the simple career reality that losses are audited more aggressively than gains.

So when the index drops and volatility rises, the demand that explodes first is often demand for *downside protection that pays quickly*. Not "protection in theory." Protection that moves today, that offsets today's P&L, that calms today's risk meeting.

That preference reshapes the skew. It makes the put wing act like beachfront property right before a hurricane: there's not much of it, everyone wants it at the same time, and the price no longer feels like a rational discount rate applied to future cash flows. It feels like a scarcity auction.

Gamma: the accelerant people feel but rarely name

One reason the elevator feels so fast is *gamma*.

Gamma is the Greek that measures how quickly an option's *delta* changes as the underlying moves. A high-gamma position means your exposure flips faster than you can blink: a small move in the underlying creates a larger change in your hedge needs.

In calm markets, dealers often carry positions that can be hedged smoothly. In turbulent markets, gamma can become a practical menace: hedging requires buying as the market rises and selling as it falls. If many participants are doing this at the same time, it can add fuel to the move.

Now connect that to skew steepening. When a crash begins, demand for puts rises. Dealers who sell those puts become short gamma and short volatility; they hedge by selling the underlying into weakness, which can deepen the selloff and reinforce the demand for more puts. It's not a conspiracy. It's a feedback loop made of rational hedging decisions that become collectively destabilizing.

This is why the market can feel like it's "falling faster than it should." The speed is partly fundamental information, partly fear, and partly the mechanical consequence of how options are hedged under stress.

The term structure flips from *savings account* to *payday loan*

In calm regimes, near-term implied volatility is often lower than longer-term implied volatility. That's the market saying, "next week is probably fine, but don't assume the year will be." It's also the market pricing the habit of selling short-dated options and buying longer-dated hedges: a kind of structural carry trade.

In panic, the market stops caring about the year. It cares about *Tuesday*.

The front end can spike so violently that term structure flips: near-term vol becomes higher than longer-term vol. That flip is the options market's version of sirens. It means immediacy has become expensive. The market is paying a premium for the next few trading sessions because it believes the distribution of outcomes over the next few days has gotten fat in both tails, and fat on the left tail especially.

The intuitive takeaway: when a crash hits, the market's fear compresses in time. It doesn't just get larger; it gets *closer*.

When the VIX print is a true scream (and when it's a weird echo)

During stress, traders often treat the VIX as the fear gauge. It's useful, but it's not a mind reader. It's a calculation based on a strip of near-dated SPX options, and during disorderly conditions, the options used in that strip can behave in ways that are more about microstructure than about a stable, consensus forecast.

One of the most educational kinds of panic is the kind that produces a *dislocation*: spot implied volatility measures jump dramatically while related instruments don't keep up in lockstep. A concrete example occurred on August 5, 2024, when spot VIX prints surged dramatically pre-market while VIX futures were far lower. The key lesson wasn't "the futures were wrong" or "spot was wrong." The lesson was that in extreme conditions, the wings—especially OTM puts—can dominate the calculation, and illiquidity can turn a theoretical index into a reflection of stressed quotes.

That matters because it reinforces a practical rule: in a crash, the price of insurance is not just economics. It is also *market plumbing*. Which strikes are liquid, where the bid-ask spreads are, how dealers mark risk, and which options are getting hit can all change the shape you think you're reading.

If you're trading skew or using it to size risk, those details can be the difference between "the market is scared" and "the market is temporarily dysfunctional."

Why ``skew steepens in a crash" is true, but incomplete

You'll hear traders say, "skew steepens in a crash." That's often directionally true because deep OTM puts get bid aggressively. But the full picture has nuance that's worth money:

- Sometimes skew steepens because the put wing rises faster than the rest.
- Sometimes skew *appears* to flatten because at-the-money vol explodes upward and the entire curve lifts, compressing the relative difference between ATM and wings.
- Sometimes the wing gets "lumpy": certain strikes become wildly rich because that's where the flow lands, while adjacent strikes lag because they're illiquid or less popular.

So rather than asking, "Did skew steepen?" ask a better question: *Where did the market choose to pay for fear?* In the belly (ATM)? In the near tail (moderately OTM puts)? In the deep tail (crash strikes)? In the front end (days to weeks)? Or across the entire horizon?

That question turns panic into a readable signal instead of a blur.

The escalator up: why volatility can sag even as prices rise

The elevator down story is emotionally satisfying, but the escalator up side is where many traders quietly lose money.

When equities drift higher, implied volatility often softens. Not because the world became objectively safer overnight, but because:

- Demand for protection relaxes when recent history feels benign.
- Realized volatility often declines in grind-up regimes (fewer large daily moves).

- Option sellers become more comfortable warehousing risk, tightening markets and compressing premiums.

This can create a seductive environment for short-vol trades: steady premium collection, stable markets, small drawdowns. The danger is that the strategy is effectively selling elevator insurance while collecting escalator premiums. It works beautifully until it doesn't, and when it doesn't, it often fails quickly.

The smirk and the surface try to warn you about this by keeping the put wing stubbornly rich even when everything looks calm. It's the market saying, "You can have your carry, but you're not getting the catastrophe for free."

What panic does to hedges: why the thing you bought may not behave like you expected

In calm markets, hedges feel like mathematics. In panic, hedges feel like personality.

A put you bought as *protection* can behave differently than you expect for a few reasons:

- *Volatility expansion*: the option can become more valuable not just because the index fell, but because implied vol rose. That's good if you're long the option, painful if you sold it.
- *Skew shift*: the specific strike you own may become much richer relative to others, which can make deep OTM protection suddenly feel brilliant.
- *Liquidity and execution*: the theoretical value may be less relevant if the market is wide and you need to transact.

> The best hedge in the world isn't helpful if you can't monetize it when you need cash.

This is where many "paper hedges" fail: they were designed in a spreadsheet that assumed continuous markets and stable relationships. Panic is the market's reminder that relationships are regime-dependent.

How to watch panic in real time without staring at a VIX chart all day

If you want a practical way to monitor panic dynamics, focus on three live comparisons rather than one headline number:

- *Spot move vs. implied move*: compare how far the index moved today to what near-dated options implied was "normal." If realized is consistently outpacing implied, fear is catching up.
- *ATM vol vs. wing vol*: if ATM is exploding but wings aren't moving as much, the market is pricing choppiness more than crash risk. If wings are inflating aggressively, the market is paying for gaps and tails.
- *Front month vs. second month*: when the very near term becomes much richer than the next maturity, immediacy is in control. That's the market paying for the next few sessions as if they're a different planet.

These comparisons are "surface-aware." They keep you from being hypnotized by a single print.

Actionable takeaway

When markets start to slide and you're deciding whether to hedge, reduce risk, or take a volatility trade, run this quick checklist before you touch anything:

1. Pull implied vols for *two maturities*: one near (about 7–14 days) and one next-out (about 30–60 days), using at-the-money options.
2. In the near maturity, compare the at-the-money IV to an OTM put IV (roughly 25-delta). Note whether the fear is building in the *belly* (ATM) or the *tail* (put wing).
3. Make one decision rule explicit: *If the front month is spiking and the put wing is getting sharply richer, treat the tape like an elevator and reduce fragility first* (position size, leverage, liquidity). If the belly is moving but wings are stable, prioritize hedges that address near-term chop rather than deep tail disasters.

The goal isn't to predict the bottom. It's to avoid making a calm-market decision with a panic-market price sheet. Panic changes the geometry of risk. If you can see the geometry changing—where vol is rising, where skew is warping, how time is being repriced—you can respond like a professional instead of reacting like a passenger who just felt the elevator cable twitch.

5.4 Beyond Delta: Vanna, Vomma, and Why Hedges Drift

A delta hedge can feel like a seatbelt. Click, tug, done. You can breathe again.

Then the market hits a pothole and your "seatbelt" turns out to be one of those retractable belts that locks when you *least* expect it. You were hedged at 10:00 a.m. By 10:07 a.m., you're inexplicably leaning into the move you thought you'd neutralized. By lunch, you're staring at a P&L line that looks like a cardiogram and wondering which part of "hedged" was supposed to be comforting.

That sensation—being hedged and still getting thrown around—is where traders discover the awkward truth: delta is only a first draft.

Delta is a snapshot, not a relationship

Delta is the simplest, most useful Greek: it tells you how much an option's price changes for a small move in the underlying. If a call has a delta of 0.40, a $1 move up in the stock is roughly a $0.40 move up in the option (all else equal). A put with delta of -0.40 moves the other way.

A delta hedge tries to cancel that first-order exposure by holding the opposite amount of underlying. It's a sensible move. It's also a snapshot. The phrase "all else equal" hides a small monster.

Because in the real market, "all else" is almost never equal. When spot moves, implied volatility tends to move. When implied volatility moves, the *shape* of implied volatility—skew and surface—often moves too. And when the shape moves, your option's delta doesn't politely stay where you left it.

So the question shifts from "What is delta?" to "What makes delta *change* when the world changes?"

That's where second-order Greeks earn their keep.

Gamma and vega: the two obvious troublemakers

Before we name the exotic-sounding ones, it helps to acknowledge the two greeks that already explain a lot of hedge drift:

- *Gamma* tells you how much delta changes when spot moves. High gamma means your delta hedge needs frequent rebalancing, especially near at-the-money and near expiration.
- *Vega* tells you how sensitive the option is to implied volatility. If implied volatility rises, long options tend to get more expensive.

If gamma is the reason your hedge needs constant babysitting, vega is the reason your hedge can be "right" on direction and still lose money. You can be delta-neutral and still be long or short implied volatility without realizing how large that exposure has become.

Now here's the twist that makes equity options feel alive: spot, volatility level, and skew are not independent characters. They're a messy friend group that travels together, especially in panic.

That interdependence is precisely what vanna and vomma are designed to describe.

Vanna: when spot moves and fear moves with it

Vanna is one of those Greeks that sounds like a designer handbag until you realize it's describing something painfully familiar.

In plain English, *vanna measures how an option's delta changes when implied volatility changes* (and equivalently, how vega changes when spot changes). It is the cross-effect between spot and volatility.

Why should you care? Because equities often have a "bad habit": when the index drops, implied volatility rises. That means a spot move is rarely just a spot move. It drags volatility along behind it like a thundercloud.

So even if you delta-hedge perfectly at the start of the day, a drop in spot can trigger a rise in implied volatility that changes your option's delta. Your hedge, which was calibrated to the old delta, becomes stale. That's hedge drift.

Here's an intuitive example without equations:

- You own a put as protection. It starts slightly out-of-the-money, with a delta that isn't very negative yet.
- The market drops a bit. Your put becomes closer to the money (gamma effect), so its delta becomes more negative.
- At the same time, implied volatility spikes because fear is rising. That volatility spike can also make the put behave

more like a real hedge right now: it fattens the tails in the model's distribution, increasing the probability-weight of finishing in-the-money.

- Net effect: the put's delta can swing more negative than you expected based on spot movement alone.

That extra "delta swing" coming from an implied vol change is the vanna story. It's the reason hedges can feel like they suddenly overreact or underreact right when you need them most.

If gamma is the steering wheel and vega is the engine, vanna is the road surface. You can be a fine driver and still skid if the pavement turns to ice.

Why vanna feels like a betrayal during selloffs

Vanna has a particularly sharp personality in equity markets because the sign of spot-vol correlation tends to be negative. Down move → vol up. Up move → vol down (often, not always).

That means your delta hedge is fighting a moving target in the exact direction you least enjoy:

- On the way down, you often become more exposed faster than your linear intuition says you should.
- On the way up, the hedge can "relax" as implied vol softens, sometimes leaving you with less protection than you thought you still had.

This is one reason crash protection is expensive: it doesn't just pay because spot falls; it pays because volatility rises at the same time. But that same feature makes hedging tricky for dealers and option sellers. Their delta risk is being pushed around by the volatility regime.

For a trader, the actionable point is simple: *delta hedging assumes delta is stable enough to hedge*. Vanna is one of the reasons it isn't.

Vomma (Volga): when volatility moves and your vega moves with it

If vanna is about the interaction between spot and volatility, vomma is about something more reflexive: *how your vega changes when implied volatility changes*.

Some desks call this *volga*; many use *vomma*. The label is less important than the sensation: you thought you had a certain vega exposure, then volatility moved and your vega exposure changed under your feet.

Vomma is "volatility of volatility" in an economic sense. It explains why a volatility spike can feel like it's feeding on itself in your risk: as implied vol rises, certain options become more sensitive to further changes in vol.

This matters most for options that are:

- Near the money (because that's where vega often concentrates), and/or
- In regimes where implied vol is moving a lot (because vomma only shows its teeth when vol actually changes).

A practical way to feel vomma is to think about an option position on a day when volatility is surging:

- If you are long options (long vega), a vol spike doesn't just give you a one-time benefit. It can make your position more sensitive to the next vol move as well, amplifying your exposure.
- If you are short options (short vega), a vol spike can make your short vega position more dangerous as it progresses. The risk you thought you were carrying can grow as the market keeps repricing fear.

That is why "short vol" strategies can look calm for months and then feel like they're blowing up in fast-forward: the risk is not always linear in volatility. Vomma is one of the reasons the risk curve bends.

The surface moves, not just the level

Most beginners think volatility moves like a thermostat: up a bit, down a bit. But real implied volatility is a surface with shape. When panic hits, the surface doesn't simply rise by 5 vol points everywhere. It can twist:

- The front end can jump more than the back end (term structure inversion).
- The put wing can inflate more than the call wing (skew steepening).
- Certain deltas can become "where the flow is," creating local ridges and kinks.

Delta hedging is already hard when *only* spot moves. It becomes conceptually wrong when spot, level, skew, and term structure are moving together. You aren't hedging one variable; you're managing a shape-shifting object.

This is why professional options risk is often discussed as "smile risk" or "surface risk," not just delta risk. You can neutralize delta and still be exposed to the market's changing preferences about *where* and *when* uncertainty lives.

A trader's story: the hedge that walked away

Picture a trader who sells a batch of index puts in a calm regime. The logic is familiar: collect premium, hedge delta, manage gamma. The position is "disciplined." The trader is not reckless; they're delta-hedged and watching the tape.

Then the index sells off. Volatility rises. Skew steepens. The put wing gets bid harder than the at-the-money.

The trader's delta hedge starts to require more frequent selling of the underlying into weakness. That's gamma at work. But there's another effect: the implied volatility jump makes the puts' deltas move faster than expected given the spot move alone. That's vanna. So each hedge rebalance is based on a delta that's changing not only because spot is falling, but also because fear is rising.

Meanwhile, the trader's vega exposure grows more sensitive as implied vol rises. That's vomma. The next volatility jump hurts more than the previous one.

This is how hedges "drift." Not because the trader forgot to hedge, but because the risk itself evolved. The hedge target moved.

The uncomfortable but useful lesson: *a delta hedge is not a promise of neutrality; it is a commitment to continuous adjustment under changing conditions*. In calm markets, the adjustments are small. In panic, they can become a full-time job.

How vanna and vomma change the way you choose strikes

This is where the topic becomes practical, not just descriptive.

Different strikes don't just have different deltas. They have different *behaviors* under stress. Deep OTM puts can behave like dead weight until they suddenly become animated by skew expansion. Near-ATM options can carry concentrated vega and gamma, making them sensitive to both spot movement and vol movement.

So if you're using options as hedges, you can choose what kind of hedge you want:

- *A hedge that responds to small moves* (more ATM, higher gamma, more immediate protection, more expensive carry).
- *A hedge that responds to regime shifts* (further OTM, more tail convexity, often cheaper day-to-day but less helpful in modest drawdowns).

Vanna and vomma are part of that decision because they tell you how the hedge will behave when implied volatility changes and when the surface warps. The "best" hedge depends on the kind of pain you're trying to survive: daily chop, fast drawdowns, gap risk, or slow grinding bear markets.

The market's smirk exists because the world is more violent on the downside. Your hedge choice is essentially a bet on *how* that violence will arrive.

A clean mental model: three knobs that turn together

Forget formulas. Think knobs.

On a stressful day in equities, three knobs often turn at once:

1. Spot moves down.
2. Implied volatility level moves up.
3. Skew gets richer on the downside (the left tail inflates).

Delta is your exposure to knob (1). Vega is your exposure to knob (2). Skew/smile risk is exposure to knob (3).

Vanna tells you that (1) and (2) are coupled: when spot moves, volatility moves, and your delta changes in response. Vomma tells you that (2) is nonlinear: as volatility moves, your sensitivity to volatility can change too.

If you only hedge delta, you're acting as if only knob (1) exists. In equities, that's like packing an umbrella for a hurricane and congratulating yourself on being prepared.

So what do professionals do differently? They hedge *dynamically* and they size for drift

There's a reason most serious volatility traders talk as much about *risk management* as they do about trade ideas. It's not because they love caution. It's because the surface can move in ways that turn a "small" position into a "large" risk quickly.

Professional practice tends to include some combination of:

- *Hedging more than one thing*: not only delta, but also some vega and some skew exposure, depending on the strategy.
- *Using structures*: spreads, collars, risk reversals, butterflies—trades that intentionally reshape which parts of the surface you're long or short.
- *Sizing with drift in mind*: assuming your hedge will worsen in a stress scenario and leaving room (capital and emotional room) to adjust.

Even if you never calculate vanna or vomma explicitly, you can trade with their reality in mind by treating hedges as living objects that will change their behavior as fear changes.

Actionable takeaway

Before you trust a "delta-neutral" label, do this simple stress-think exercise on any option position that matters to your account:

1. Assume the underlying drops 2% *and* implied volatility rises (pick a realistic jump for the market you trade—equities often do this together).
2. Ask: *Would my delta hedge still be correct, or would the option's delta likely become more negative (puts) / less positive (calls) because implied vol rose?* That question is vanna in plain English.

3. Ask: *If implied volatility rises again after the first jump, would my sensitivity to that second jump be larger than it was before?* That question is vomma in plain English.

If either answer is "yes," size smaller than your calm-market intuition suggests, and plan your re-hedge points in advance. Delta hedging isn't broken—it's just incomplete. In equity volatility, the market doesn't move like a straight line. It moves like a shape that reacts to itself. Vanna and vomma are the names for that self-reaction, and respecting them is the difference between a hedge that helps and a hedge that quietly walks away when you need it most.

5.5 Trading the Shape: Risk Reversals and Butterflies

A lot of option trading is secretly just arguing with a menu.

Not the menu of stocks you like or don't like. The menu of *prices for fear*, arranged by strike and expiration. One item is "panic right now." Another is "catastrophe insurance if the floor collapses." Another is "I want upside, but I don't want to pay for it."

Most beginners order the obvious thing: a call because they're bullish, a put because they're bearish. Professionals often order something else: they buy what looks underpriced on the surface and sell what looks overpriced—even if they don't have a strong view on direction.

That's what "trading the shape" means. You're not just trading whether the market goes up or down. You're trading how the market is *pricing* up versus down, near-term versus long-term, wings versus belly.

Two structures show up again and again because they map cleanly to the two most important features of the equity vol surface:

- *Risk reversals* are a bet on *skew* (the smirk).
- *Butterflies* are a bet on *curvature* (how fat the wings are relative to at-the-money).

They're also the language dealers and institutional traders use to quote "shape" without talking in paragraphs. If you can understand a risk reversal and a butterfly, you can read a huge amount of the options market's subtext.

First, a reality check: you can't trade shape without respecting tail risk

Equity skew isn't a cute quirk; it's a scar. It exists because the downside has behaved like an elevator often enough that the market refuses to sell the insurance cheaply.

So when you see a structure that seems to "harvest" that premium, don't treat it like free money. Treat it like you're taking the other side of a crowd's worst fear. That can be smart, but only if you know exactly what you're short.

The right mindset is: *shape trades are relative value trades with absolute consequences.* You're comparing rich and cheap parts of the surface, but you're still exposed to the world.

Risk reversals: selling fear to buy hope (and why it's never neutral)

A *risk reversal* is the simplest shape trade that feels like a magic trick: you buy an option on one side and finance it by selling an option on the other side.

In equity indices, the classic "recovery" risk reversal is:

- *Sell an OTM put* (expensive because of skew).
- *Buy an OTM call* (cheaper relative to the put).

Done at comparable deltas (for example, short 25-delta put and long 25-delta call), this structure is essentially saying: "Downside insurance is overpriced relative to upside participation."

That's a skew view, not just a directional view. If the put wing is extremely rich, you may be able to buy a decent upside call for little net premium, sometimes close to zero-cost.

It feels like you've outsmarted the market. What you've actually done is choose a very specific personality for your P&L.

What a risk reversal is *really* long and short

A risk reversal is not a "cheap call." It's a package with hidden exposures:

- You are *short left-tail convexity*. The short put is the part that hurts in an elevator-down market.
- You are *long upside convexity*. The long call pays if the market rallies.

- You are expressing a view that the skew is too steep: that the market is overpaying for downside fear relative to upside.

You can be right about skew and still lose money if the underlying drops hard enough. The short put doesn't care that you got a "cheap" call. It cares that you sold insurance.

This is why risk reversals are loved by traders and hated by risk managers: they can look elegant on entry and brutal in a tail event.

When a risk reversal makes sense

A risk reversal tends to be most defensible when the world you're short (a sharp crash) is less likely *for the next few weeks* than the market is charging you for, and when you have a plan for the scenario where you're wrong.

In practice, that often means:

- Skew is unusually steep (put wing is very rich relative to calls).
- You have a reason to believe near-term crash risk is overpriced *relative* to upside (for example, positioning is extremely defensive, or a known catalyst has passed).
- You can define risk limits on the short put leg (through sizing, spreads, or stop/roll rules).

The goal isn't to declare "no crash." The goal is to say: "The market is charging an unusually high premium for crash insurance, and I'm willing to be a seller *with limits*."

Make it survivable: the risk-defined risk reversal

If you like the skew view but hate undefined disaster exposure, there's a simple modification: cap the downside.

Instead of selling a naked put, sell a put spread:

- Sell an OTM put.
- Buy a further OTM put (same expiration).
- Use the credit to finance the call purchase.

You've turned "sell fear to buy hope" into "sell *some* fear to buy hope." You may not get the call as cheap, but you've replaced catastrophic loss potential with a known worst-case.

This matters because skew is most seductive right before it matters most. The premium looks juiciest precisely when the crowd is most nervous for a reason.

Butterflies: betting that the market overpriced the wings (or the belly)

If risk reversals are about left-versus-right, butterflies are about wings-versus-belly.

A *butterfly* (often just "fly") is typically built with three strikes in the same expiration:

- Buy one lower strike option.
- Sell two middle strike options.
- Buy one higher strike option.

Calls or puts can be used; the structure is the same. The payoff is shaped like a tent: it benefits if the underlying finishes near the middle strike, and it loses if the underlying finishes far away in either direction.

That sounds like a boring "range bet," but the more interesting interpretation is about implied volatility shape:

A butterfly is a trade on curvature. It's you saying: "The market is charging too much (or too little) for the wings relative to the center."

In equity indices, wings are often rich because skew and tail demand keep them bid. Sometimes they get *too* rich. Sometimes the belly gets too rich because the market is bracing for choppy, mean-reverting noise that doesn't arrive. Butterflies let you take a position on that relative pricing.

Long fly versus short fly: which part of the surface you're buying

Two quick translations make butterflies less mysterious:

- A *long butterfly* (buy wings, sell belly) tends to like a world where the market doesn't move much and/or where implied volatility comes down. You're often long options on the wings but short more option in the middle, so you're implicitly saying: "The middle is too expensive relative to the wings," or "realized will be tame."
- A *short butterfly* (sell wings, buy belly) is often a way to sell wing richness. You're saying: "The wings are overpriced relative to ATM." In equity indices, this can be a way to monetize extreme tail bidding—but it can be dangerous if the tail actually shows up.

The naming can be confusing because people describe flies in different quoting conventions (delta flies, strike flies, call flies, put flies). Don't get hung up on the vocabulary. Get hung up on the economics: *what are you selling, and where does it hurt?*

Why traders talk in ``ATM + RR + Fly"

On many professional desks, implied volatility is effectively decomposed into building blocks:

- *ATM volatility*: the overall level.
- *Risk reversal* (*RR*): the tilt, or skew (put versus call richness).
- *Butterfly* (*Fly*): the curvature, or how expensive wings are relative to ATM.

This is not just trader poetry. It's how people manage surface risk. If your book is hurt by skew moves, you hedge with risk reversals. If your book is hurt by curvature moves (wings inflating or deflating relative to ATM), you hedge with flies.

That's the practical connection to hedging drift: delta hedges the first-order spot move, but it doesn't hedge shape. RR and fly are ways to hedge and trade *shape* directly.

A concrete scenario: ``panic premium" in the put wing

Consider a market that has recently had a scare: a sharp selloff, a volatility spike, headlines that made everyone check their accounts twice. Even after the market stabilizes, downside skew can remain elevated. People don't forget quickly.

In that regime, deep OTM puts can be rich relative to everything else. The surface is basically saying: "I'm calmer today, but don't you dare take away my disaster insurance."

A trader who believes the immediate panic is fading might want to express: "Those far OTM puts are too expensive relative to nearer strikes or relative to ATM."

There are multiple ways to do that, but butterflies are a classic tool because they can target wing richness more precisely than a blunt short put.

For example, a put butterfly can be constructed so that your exposure is concentrated in a region of strikes where the implied volatility "ridge" looks exaggerated. You're not betting on up or down in the usual sense; you're betting on a specific part of the surface relaxing.

If you're right, the fly benefits from the repricing of that curvature. If you're wrong and the market drops hard again, you learn why the ridge existed.

Why ``shape trading'' is often just ``risk management with better posture''

A common misconception is that surface trading is something you do instead of directional trading. In reality, many shape trades are directional trades that have been made *more honest*.

If you want upside exposure but don't want to pay full premium, a risk reversal can give you a way to express that while acknowledging what you're financing it with.

If you believe realized volatility will be lower than implied but you're nervous about a tail event, a butterfly can let you express "vol is too expensive" in a more nuanced way than simply shorting straddles.

Shape trading forces you to say which part of the surface you think is wrong. That clarity is useful even when you're wrong, because it tells you *what* you were wrong about: level, skew, or curvature.

Practical guardrails: how to keep shape trades from turning into horror stories

These structures are powerful, and they come with the kind of risks that don't announce themselves with a polite email.

A few guardrails that actually help:

- *Don't confuse "financed" with "free."* If you sold a put to buy a call, you paid with tail risk.
- *Use risk-defined variants when possible.* Put spreads instead of naked puts. Wing spreads instead of naked wing shorts. Know your worst-case.
- *Match your thesis to your tenor.* If you're trading a short-term skew dislocation, don't express it in a six-month structure that will bleed and morph.
- *Plan the unwind.* Shape trades often look best on entry. Decide in advance what success looks like (a skew normalization, a fly cheapening, a certain P&L) and what failure looks like (spot move threshold, vol spike, skew blowout).

Those aren't moral lessons. They're survival lessons. Shape trading is often about selling what's crowded. Crowded trades can stay crowded longer than you want—and then they can move violently.

Actionable takeaway

The next time you look at an option chain and feel tempted to trade direction, force yourself to answer one extra question:

"Am I actually trading direction, or am I trading a mispricing in the surface?"

Then do this quick diagnostic:

1. Compare 25-delta put IV to 25-delta call IV in the same expiration. If the gap feels extreme, you're looking at a skew problem—a risk reversal is the natural instrument to express or hedge that view.
2. Compare wing IVs (say, 10–25 delta options) to ATM IV in the same expiration. If wings look absurdly rich (or surprisingly cheap) relative to ATM, you're looking at a curvature problem—a butterfly is the natural instrument to express or hedge that view.
3. Choose a *risk-defined* version by default unless you have a very explicit reason not to. If the trade looks too elegant to require guardrails, it probably needs them most.

The surface is a pricing map. Risk reversals and butterflies are the tools that let you trade the map itself, not just the direction of the road.

5.6 The Cost of Catastrophe Insurance

The most dangerous thing about a deep out-of-the-money put is how good it looks in hindsight.

After a crash, you can pull up a chart of some "disaster hedge" that turned $10,000 into $400,000 and feel a clean, righteous anger at your past self for not buying it. The trade looks like moral clarity: the world is chaotic, risk is real, and you were smart enough to prepare.

Then you run the clock forward from any random calm month and actually *hold* that hedge.

The option decays. It expires worthless. You buy another. That one decays. You do it again. After a year of paying premiums, you start to tell yourself you're "disciplined" because you're consistent. After two years, you start to feel like you're sponsoring the market's liquidity providers.

That's the economics of catastrophe insurance: it's supposed to feel expensive most of the time. If it didn't, the people selling it would be volunteering for misery.

Tail hedging is a real product, not a clever trick

A deep OTM put is not a *trade idea*. It's a product with a purpose: it is meant to pay during periods when everything else in the portfolio is getting punched at the same time.

When equities crash, three things tend to happen together:

- Spot drops fast.
- Implied volatility rises.

- Skew and the left tail often get richer (puts become more expensive even before you buy them).

A put hedge is one of the few instruments that can benefit from all three simultaneously. That's why it's coveted. That's also why it carries a persistent premium.

The premium is the market's way of saying: "If you want something that performs when everyone is desperate, you will pay for it while you're calm."

The unromantic math: negative carry and bleed

Tail protection has a cost that shows up as *negative carry*. Traders call it *bleed*: the slow, regular loss from paying premium for something that usually doesn't pay out.

If you buy a put and nothing dramatic happens, time passes and the option's value tends to fall. That's not a flaw in the system; it's the system working as designed. You bought a decaying asset. You paid for the right to be protected during a window of time. If the disaster didn't show up during that window, the protection expires.

The important part is what happens when you do it *repeatedly*:

- A single put can be a smart hedge.
- A program of always owning deep OTM puts is an *insurance business*—and you are the customer, not the underwriter.

Insurance customers shouldn't expect positive expected value on the premium payments. They should expect *utility*: fewer sleepless nights, lower drawdowns, the ability to stay invested through a crisis instead of selling at the bottom.

That's a valid goal. Just don't confuse it with a free lunch.

Why the market overcharges you (on average) and calls it ``fair"

There's a recurring pattern in equity index options: implied volatility tends to run higher than subsequent realized volatility on average. In plain terms, option buyers often pay more than the future ends up "justifying" if you measure justification as the actual volatility that occurs.

This difference is often described as a *volatility risk premium*: the extra compensation option sellers demand for taking the other side. It's not charity. Sellers are warehousing a risk that is unpleasant in precisely the way humans hate: it arrives in clusters, it arrives when liquidity is worst, and it arrives when correlated assets fail together.

Deep OTM puts concentrate that unpleasantness. You are asking someone to sell you protection against the kind of day that can break a market maker's month, a fund's year, or a career. The market responds by charging you a premium that can feel insultingly high.

That insult is the point. The market is discouraging casual buyers. It is pricing scarcity of willingness, not just probability.

The ``flood insurance" problem: the time you want to buy is the time it's least available

If tail hedging were only about logic, everyone would buy protection when it's cheap and relax when it's expensive. In practice, people do the opposite, because the desire to hedge is emotional and often arrives late.

The pattern is painfully human:

- Calm regime: protection feels unnecessary; buying it feels like admitting you're afraid; the bleed feels annoying.
- Early selloff: protection suddenly feels wise; you start pricing it out.
- Panic: the quotes widen, the wing gets bid, and the exact put you want becomes *the most expensive thing on the shelf*.

This is why tail hedging is hardest precisely when it feels most urgent. When fear is loud, you don't just pay more. You sometimes can't get clean execution at all.

In extreme conditions, even the "fear thermometer" style indices can produce prints that are influenced by microstructure and stressed quotes. That's not academic trivia. It's the lived experience of trying to buy catastrophe insurance while the catastrophe is beginning.

So the only rational time to buy tail insurance is when you *don't* feel like buying it. Unfortunately, humans are not machines. They're story-driven animals with performance reviews.

The hidden cost: opportunity loss and behavior

Tail hedging costs money, but the bigger cost can be psychological: it changes how you behave.

If you're always paying premium, you may feel permission to take more equity risk elsewhere: more leverage, less diversification, tighter concentration. That can be rational if the hedge is sized correctly and truly offsets the worst outcomes.

But it can also become a subtle self-deception: "I'm protected, so I can ignore risk." Then the hedge bleeds for years, you reduce its size out of frustration, and the crash arrives right after you stop paying.

The most common failure mode in tail hedging isn't math. It's endurance.

Tail hedging is not a one-time purchase. It's a program. Programs require a reason you can keep believing on boring days.

Deep OTM puts: why ``cheap" is often a mirage

A deep OTM put can look "cheap" because its premium is small in dollars. But small premium can hide two realities:

- *It may be too far away to help.* If the strike is so low that a typical correction doesn't reach it, the hedge won't pay when you actually feel pain.
- *It may require a very specific kind of crash.* Deep OTM puts often need a fast, violent move (and a vol spike) to become meaningful before expiration.

There's a reason professional hedgers spend so much time choosing strikes and expirations. The hedge needs to be close enough to activate in realistic stress, long enough to survive noise, and cheap enough not to become a permanent performance anchor.

A $0.30 put that never pays is not "cheap." It's a subscription.

So when do tails actually pay?

Tail hedging can pay in two ways, and it's helpful to separate them:

- *Payment in P&L*: the hedge makes money during a crash, offsetting losses elsewhere.
- *Payment in behavior*: the hedge keeps you invested and prevents the worst decision in finance: panic-selling into the hole.

If you're hoping for consistent P&L outperformance from always owning deep OTM puts, you're fighting the insurance business model. The market is engineered so the seller usually wins in the long run. That's why the seller exists.

If you're aiming for behavior protection—a smoother ride, fewer forced liquidations, the ability to rebalance into chaos instead of away from it—the hedge can be worth the bleed.

The best tail hedgers tend to be honest about what they're buying. They aren't buying a lottery ticket. They're buying the ability to make rational decisions on irrational days.

Better questions than ``Should I buy puts?"

The right question isn't "Should I buy tail protection?" The right question is "What kind of failure am I trying to prevent?"

Here are a few concrete failure modes tail protection can address:

- *Margin failure*: you don't want to be forced to liquidate at the worst time.
- *Mandate failure*: you can't tolerate a drawdown beyond a certain point without violating a risk policy.
- *Behavioral failure*: you know you will not hold through a 30% drawdown without some offset.
- *Liquidity failure*: you need cash during stress when selling assets is costly.

Once you know the failure mode, the hedge can be designed to address it. That might mean deep OTM puts, but it might also mean put spreads, collars, longer-dated protection, or simply a less fragile portfolio. Options are powerful, but the best hedge is often the one that prevents you from needing to hedge in the first place.

A practical framework: make the bleed explicit

If you decide to run a tail-hedge program, treat it like an insurance budget, not like a trade.

Pick a number you can live with: for example, "I'm willing to spend X basis points per year to reduce catastrophic drawdown risk." Then design the hedge to spend roughly that budget under normal conditions.

This matters because the worst tail hedging decisions happen when the cost is not explicitly owned. If you don't choose the bleed, the bleed will choose you. You'll resent it, cut it at the wrong time, or size it inconsistently.

Consistency is not about virtue; it's about avoiding the most predictable mistake: buying expensive protection in fear and canceling cheap protection in calm.

The uncomfortable truth: the best tail hedge is often boring

Deep OTM puts are dramatic. They're cinematic. They produce the screenshots people share.

But the best risk management often looks boring from the outside:

- Less leverage than your ego wants.
- More liquidity than your optimism thinks you need.
- Hedges that activate in the range where you actually start making bad decisions, not only in apocalyptic scenarios.

If your tail hedge requires the world to end to be worth it, it will fail you most of the time. If your hedge is designed to keep you functional during stress, it will look less sexy and work more often.

Actionable takeaway

If you're considering buying deep OTM puts as catastrophe insurance, do this before you place the order:

1. Write down the specific failure you're preventing (margin call, mandate breach, behavioral capitulation, liquidity need). If you can't name it, you're shopping emotionally.
2. Choose an annual "insurance budget" in basis points of your portfolio that you can tolerate bleeding through calm markets. Make it explicit.
3. Back into strikes and expirations that actually address the failure mode: close enough to matter, long enough to survive noise, and sized so you can keep paying the premium without rage-quitting.

Catastrophe insurance is not supposed to be a positive expectancy trade. It's supposed to be a negative-carry asset that buys you something priceless: the ability to stay solvent, and stay rational, when the market stops being either.

Chapter 6

The Playbook: Strategies for Every Regime

If the previous chapters were the flight manual, this is where we actually grab the stick. The market isn't a monolith; it has seasons, moods, and violent tantrums. Most traders fail because they use a sunny-day strategy in a thunderstorm. In this chapter, we open up the toolbox. We'll look at how to be the 'House' by selling premium, how to hunt for Black Swans without going broke waiting for them, and how to exploit the quirky math of the VIX curve. Whether you want to grind out monthly income or position yourself for the next crash, there is a specific play for that-and here is how you execute it without blowing up your account.

6.1 Selling Volatility: Harvesting Premium

The first time you sell an option for premium, it can feel like you discovered a cheat code. Money arrives in your account *today*. Your brain files it under *income*. You start doing the little math tricks that humans love: "If I can make $300 a week, that's $15,600 a year..." And because the market often behaves itself for long stretches, you get a run of trades that seem to confirm the story: sell premium, let time pass, buy it back cheaper (or let it expire), repeat.

That feeling is not an accident. Options are the insurance market of finance, and like most insurance markets, the average buyer is willing to overpay for peace of mind. People don't buy homeowners insurance because the expected value is positive; they buy it because the downside is life-altering. Options work the same way: plenty of market participants will pay extra to avoid the stomach-drop of watching a portfolio vaporize during a violent selloff. That extra is the *fear premium.*

In options language, the fear premium is usually discussed as the *Volatility Risk Premium* (VRP): implied volatility (the volatility embedded in option prices) tends to be higher than the volatility the market actually realizes afterward. That gap is why selling options can feel like owning a small toll booth on human anxiety. You collect a steady drip of nickels and dimes from people who want protection, plus a little extra because protection is most demanded exactly when it's hardest to provide.

The trap is that the insurance company's business model comes with an insurance company's problem: catastrophe risk. Most days are quiet. A few days try to bankrupt you.

The house edge (and why it exists even in a competitive market)

If markets are efficient, why would option buyers systematically overpay? The answer isn't "because everyone is dumb." It's because option *sellers* are taking a very particular kind of risk that investors hate: losses that arrive fast, at the same time as everything else is going wrong.

When equities sell off hard, correlation goes to one, liquidity gets thin, spreads widen, and margin demands rise. A short option position doesn't just lose money; it tends to lose money when it's hardest to raise cash and when your other positions are probably bleeding too. That's not normal risk. That's *funding and timing* risk—the kind that forces liquidations at the worst possible moment.

So the market demands a premium for carrying that risk. In plain English: if you're going to be the insurer, the market insists you charge enough that it's worth waking up at 3 a.m. to handle the occasional fire.

There's also a mechanical reason VRP persists: many large institutions are structural buyers. Pension funds, endowments, risk-parity funds, and systematic strategies often have mandates that translate into persistent demand for protection or convexity. Even when nobody is panicking, somebody is hedging. And when a steady buyer meets a seller who requires compensation for catastrophe timing, prices lean upward.

The point isn't that selling volatility is "free money." The point is that it has a business logic. You're not betting on a coin flip. You're underwriting risk for a premium because the risk is nasty in precisely the way humans and institutions most dislike.

The options seller's personality test: do you enjoy being right 9 times out of 10?

Short volatility trades often win with high frequency. That's the seduction. Most option contracts expire worthless. Most market days are not crisis days. Most big scary headlines don't turn into limit-down cascades. If you structure your trades with defined risk and sane sizing, you'll often see a streak of small wins.

That win rate is not the same as a good strategy.

Short vol tends to have a return profile that looks like this: many small gains, and occasional large losses. Statisticians call this *negative skew*. Traders call it "picking up pennies in front of a steamroller." The pennies are real. The steamroller is also real. The career-ending mistake is forgetting that the steamroller doesn't have to show up often to dominate your lifetime P&L.

Here's the uncomfortable truth: a lot of "successful" short-vol traders are simply traders who haven't met their steamroller yet.

Why the steamroller is gamma (and why gamma doesn't care about your opinion)

To understand why short volatility blows up, you only need one Greek: *gamma*. Gamma measures how quickly an option's delta changes as the underlying price moves. High gamma means delta moves fast, which means your position's directional exposure can balloon right when the market is moving fastest.

If you sell options, you are typically *short gamma*. Translation: when the market moves against you, your position tends to become *more* exposed, not less. A short put gets more delta-negative as the stock falls. So you're losing money, and your position is leaning harder into the fall. That's why crashes feel like your risk is multiplying.

This is the opposite of what most people expect from "income." Income feels like a salary: steady, predictable, something you can spend. Short option premium is more like running a nightclub: most nights you make money; once in a while someone drives a car through the front window.

Gamma risk is also why short vol strategies can look brilliant in a backtest that doesn't model realistic execution. When the market gaps, you don't get to adjust smoothly. When volatility spikes, bid/ask spreads widen. When your broker raises margin requirements, your best theoretical hedge becomes an expensive practical hedge.

If you sell premium, you're not just selling odds. You're selling *liquidity* to the market in its worst moments.

Two ways to sell premium without lying to yourself

There are countless short-vol structures, but two show up again and again because they make the risk visible:

- *Credit spreads*: sell an option and buy another farther out-of-the-money (same expiration, same type) to cap the maximum loss.
- *Iron condors*: a put credit spread plus a call credit spread, usually placed around the current price to bet on a range and collect premium from both sides.

A credit spread is the simplest confession a short-vol trader can make: "I want to sell fear, but I'm not willing to risk my entire account on a gap." An iron condor is the more ambitious confession: "I want to sell fear on both sides because I think the market is overpricing movement in general."

Both can be intelligent. Both can be disastrous if sized like a paycheck instead of an insurance book.

The key is that these are not bets on direction. They are bets on *realized movement* relative to what's priced. If the market moves less than implied volatility suggested, you get paid. If it moves more—especially quickly—you pay.

A mental model that keeps you honest: you're selling a monthly subscription to disaster

One reason short vol traders overtrade is that each premium sale feels like a discrete opportunity. It's not. It's closer to selling a subscription. You're repeatedly taking on the same kind of exposure to the same kind of event: sudden, violent repricing.

Imagine you sold flood insurance in a coastal town. Month after month, you'd collect premiums, feel clever, and start to treat the income as normal. Then a hurricane hits. Your past year of "profits" was never profit; it was just *unearned premium* waiting for the right weather.

In markets, the hurricane is a volatility regime change. It doesn't have to be 2008. It can be an unexpected CPI print, a surprise central bank pivot, an ETF unwind, a geopolitical escalation, or a liquidity event that makes perfectly normal hedging temporarily impossible.

That's why professional vol sellers talk less about "being right" and more about *surviving*.

The return distribution: pennies are frequent, steamrollers are rare

Most investors intuitively think of returns like a bell curve: lots of medium outcomes, few extremes. That's closer to how long stock behaves over short horizons (still not perfectly bell-shaped, but the intuition holds). Short-vol returns are different. They're lopsided: clustered small wins with a long, ugly tail of rare crashes.

Place this picture in your mind before you place your first trade:

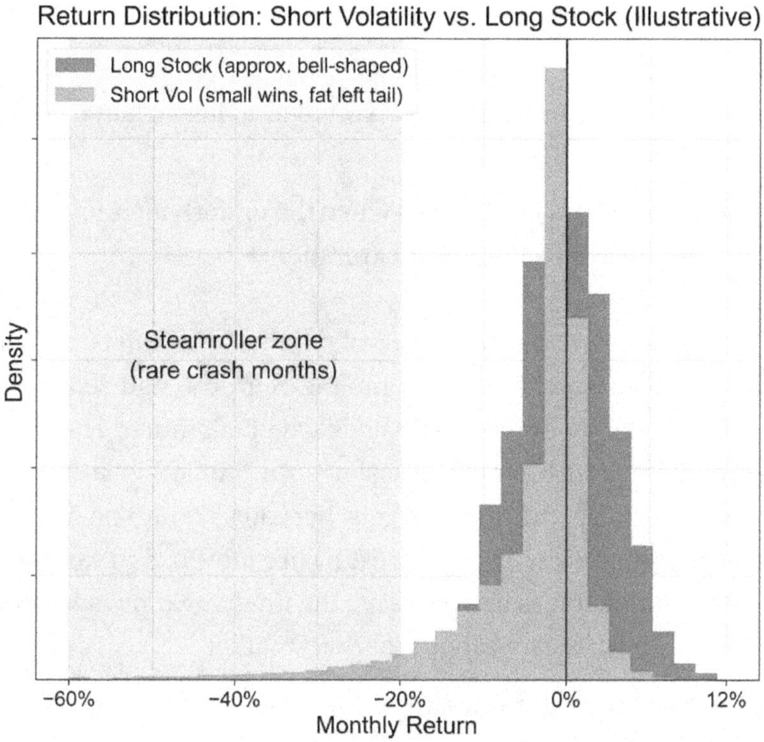

Note: The data presented in this figure is simulated and for illustrative purposes only.

That left tail is the bill you eventually have to pay for all those tidy months. The market is not promising you steady income; it's offering you a trade: collect premium most of the time, then write a check during stress. The entire craft of selling volatility is learning how to keep that check from being fatal.

The practical mechanics: what you're actually selling

When you sell an option, you sell a bundle of exposures. Even if you don't use the Greeks day-to-day, it helps to know what levers are driving the P&L:

- *Theta* (time decay): you benefit as time passes, all else equal.
- *Vega* (implied volatility): you tend to lose when implied volatility rises.
- *Gamma*: you tend to lose when the underlying moves fast and far, especially near expiration.

This explains a common short-vol paradox: you can be "right" about direction and still lose money. Suppose you sell a put spread because you think the stock won't fall much. The stock drifts down a little—not a disaster—but implied volatility explodes because the market gets nervous. Your spread can widen dramatically even if price hasn't moved that far. Your mark-to-market loss arrives *before* the final outcome is known, and your broker cares about mark-to-market.

Selling volatility is a trade in *paths*, not just endpoints.

Credit spreads: the disciplined version of short premium

A credit spread is two options with the same expiration:

- You sell the nearer strike (collect premium).
- You buy a farther strike (pay premium) to cap losses.

A put credit spread might be: sell the 95 put, buy the 90 put. The credit you collect is your maximum profit. Your maximum loss is the strike width minus the credit, times the contract multiplier.

The good news: you can't lose infinity. The bad news: your maximum loss is often many multiples of your maximum gain.

If you collect $1.00 on a $5-wide spread, you risk $4.00 to make $1.00. You are literally running an insurance book: small premiums against larger claims.

That's not inherently bad. Insurance companies do it. The trick is that insurance companies diversify, size conservatively, and stay solvent through the catastrophe. The amateur mistake is to sell one big policy (one big spread) on one neighborhood (one underlying) right before the hurricane season (earnings, CPI, Fed meetings) and call it a strategy.

Iron condors: selling movement itself

An iron condor adds a call spread on top of a put spread. You're basically saying: "The market is pricing too much movement in either direction. I think the realized move will be smaller than implied."

When iron condors work, they feel smooth. The underlying chops around, time decays, both spreads shrink, and you buy them back for pennies. That smoothness is precisely why iron condors can seduce traders into oversizing.

When iron condors fail, they fail in a very specific way: the market trends. Not wiggles. *Trends*. One side goes deep in the money, implied volatility often rises, and the position becomes a directional anchor tied to your leg width and your willingness to adjust.

If you're going to sell condors, you should be able to answer a simple question without blinking: *What will you do if the underlying touches your short strike quickly?* Not "I'll manage it."

Specific actions: close, roll, reduce, hedge, or accept max loss. Vagueness is expensive.

The hidden enemy: correlation and ``everything blows up together''

A short-vol trader's worst surprise is not that one stock moved. It's that *all* stocks moved, *together*, *fast*. That's why diversification by ticker often fails when you need it most. In quiet markets, your short premium positions look diversified. In a shock, they can become one giant bet on index volatility.

That's the emotional whiplash of the strategy. You can spend months feeling like a careful risk manager—different sectors, different expirations, different structures—and then a macro event turns your portfolio into a single exposure: short market convexity.

This is why many professional vol sellers anchor risk at the index level (SPX, broad ETFs) even when they trade single names. The index tells you what regime you're in. Single-name "diversification" tells you bedtime stories.

The only sizing rule that matters: survive the steamroller

Most blow-ups in short vol are not about being "wrong." They're about being too big.

Sizing is the difference between a drawdown and a death spiral. When volatility spikes, you can face:

- expanding option prices (vega),
- accelerating delta exposure (gamma),

- widening spreads (execution),
- higher margin requirements (funding),
- and forced de-risking (psychology + broker).

So here is a conservative, practical framework that keeps you alive even when your ego wants to swing bigger:

- Treat each short-vol position as if it *will* hit its worst-case loss at least once in your trading life.
- Limit total worst-case loss across all open short-vol positions to an amount you can take *without changing your life*. For many retail accounts, that's single-digit percentages of the account, not double digits.
- Keep dry powder. If you are fully margined in calm markets, you are volunteering for forced liquidation in stress markets.

If that sounds too cautious, good. Short volatility punishes bravado and rewards boring solvency.

Trade selection: when selling premium is rational (and when it's just dopamine)

Short vol works best when two conditions are true:

- implied volatility is elevated relative to recent realized movement (you're being paid),
- and the environment is not obviously fragile (you're not selling hurricane insurance during a hurricane watch).

The second condition is the one traders ignore because it's harder to quantify. They look at a high IV rank or a juicy credit and feel "paid." But the market doesn't pay you extra because it's feeling generous. It pays you extra because it's nervous.

Sometimes the market is nervous for silly reasons. Sometimes it's nervous because something is genuinely unstable.

Your job isn't to predict the news. Your job is to recognize when the market's nervousness is *cheap* versus *expensive*. Cheap nervousness is chatter and minor events priced like disasters. Expensive nervousness is when liquidity is fragile, correlations are rising, and macro catalysts are stacked back-to-back. In expensive nervousness, the premium is high because the steamroller is closer than usual.

A trader who sells premium through every regime is not a disciplined premium harvester. They're a person who hasn't learned to say no.

Execution details that separate pros from ``it looked fine on the app"

A few operational truths matter more for short vol than for almost any other style:

Defined risk is not the same as small risk. A $5-wide credit spread is "defined risk" but it can still be large relative to your account if you stack too many. Defined risk just means you know the maximum loss, not that you can afford it.

Near-expiration gamma is a different animal. Short-dated options can look like free money because theta decay accelerates. But gamma also accelerates. Selling premium with just a few days to go can turn your position into a hair-trigger directional exposure. Many blow-ups happen in the final week, not because the trader was unlucky, but because they were short the most explosive version of gamma.

Liquidity is a strategy variable. In calm markets, you can usually get fills near mid. In stress, the spread *is* your edge evaporating. If your plan requires surgical adjustments during a panic, you're building a strategy on a day that won't exist when you need it.

Assignment risk is real. If you trade American-style equity or ETF options, short in-the-money options can be assigned early, especially around dividends or near expiration. Many traders learn this the hard way: they wake up long or short shares they never intended to own, with margin and risk profiles they did not plan for.

Insurance companies don't "forget" the fine print. Neither can you.

A simple way to think about exits: close winners early, close losers earlier

A short premium trade has asymmetric risk: limited profit, larger loss. That asymmetry suggests a counterintuitive discipline:

- Take profits before you "earn" the last scraps of premium.
- Cut risk quickly when the trade changes character.

If you sold a spread for $1.00 and it's now worth $0.20, you've captured 80% of the potential profit while leaving yourself exposed to the exact same tail event for the remaining 20%. Many traders stay in because they hate leaving money on the table. But that last 20% is often the most dangerous, because time is short and gamma is waking up.

Similarly, a loser that is widening rapidly is not always a candidate for "patience." Sometimes it's telling you the market has entered the regime where your strategy bleeds. The best short-vol traders are not the ones who never take losses. They're the ones who don't turn manageable losses into existential ones.

The real objective: monetize VRP without becoming a volatility charity

Selling volatility can be a legitimate, repeatable way to harvest the fear premium. It can also be a slow-motion path to ruin if you treat premium like yield and ignore the embedded catastrophe exposure.

The honest goal is not "collect premium every week." It's: collect premium when you are paid enough for the risk *and* sized such that the inevitable ugly month is a scar, not an obituary.

Actionable takeaway

Before placing the next short-premium trade, write down two numbers and refuse to trade without them:

1. *Max loss you can tolerate on this position* (in dollars and as a percent of account), assuming the worst case happens quickly.

2. *Profit level where you will exit early* (for example, buying back the spread when you've captured 60–80% of the credit), so you're not risking a steamroller for leftover change.

If you can't state those two numbers calmly, you're not harvesting premium. You're renting out your account as the market's shock absorber.

6.2 Buying Volatility: Catching the Crash

A long-volatility trade is the financial equivalent of paying for a fire extinguisher and then staring at it every month, wondering if you're the sucker. It sits there, quietly depreciating. It doesn't throw off income. It doesn't feel productive. Most of the time it looks like you bought something that doesn't work.

Then the smoke alarm goes off.

When markets break, they don't break politely. Prices gap. Correlations snap tight. Liquidity thins out exactly when everyone wants it. The same headlines that would have been shrugged off in a calm regime suddenly hit like a brick because everyone is already leaning the same way. That's the moment long volatility is designed for: the moment when you're not trying to be *right* about direction so much as you're trying to be *paid* for speed and violence.

Buying volatility means owning options in a way that benefits from big moves and volatility spikes. It's the inverse business model of selling premium. Instead of collecting the fear premium, you pay it—over and over—while you wait for the day the receipt finally matters.

The cruel math of being early

The market has a special talent for making crash protection feel embarrassing. You buy puts, the market grinds higher, and the puts bleed like a slow leak. You roll them, you bleed again. If you're unlucky, you'll do this through an entire bull run and conclude that long vol is a scam.

It's not a scam. It's a timing problem.

Long volatility has *negative carry*. Carry is what your position earns (or loses) just by existing. A bond pays carry via coupons. A short option position often earns carry via time decay (theta). A long option position typically *pays* carry because time works against it. Every day that passes without the move you need is a day your option gets a little closer to expiring worthless.

That grind is not a bug. It's the price of convexity.

Convexity is the magic trick long vol is buying: a payoff that accelerates as the market moves further. It's not linear like stock exposure; it's curved. Small moves can hurt (because theta dominates), but large moves can overwhelm the bleed. This is why crash protection can look awful for months and then look genius in a single week.

The difficulty is emotional and practical: most traders can tolerate being wrong. Far fewer can tolerate being *right too soon*.

Why ``I think the market will go down" is not a long-vol thesis

A common rookie move is to buy puts because you're bearish. Sometimes that works, but it confuses two different ideas:

- *Direction*: will the underlying go down?
- *Volatility*: will the underlying move more than the option market is pricing in, and/or will implied volatility rise?

You can be directionally correct and still lose money on a long option if the move is too slow, too small, or comes with an implied-volatility collapse. Traders call that a *vol crush*: the market relaxes, implied volatility drops, and option prices deflate even if the underlying moved in your favor.

This is one reason earnings trades humble people. A stock can drop after earnings and the put buyer still loses because the options were priced for a larger move and pumped with pre-event implied volatility. You bought an umbrella during a hurricane warning, it rained, and the umbrella still wasn't worth what you paid.

Long-vol trading is less about being bearish and more about having a view on *speed*, *magnitude*, and *surprise*.

The three enemies: theta, the calendar, and your own attention span

Long vol fails in three predictable ways.

Theta: the daily tax. Even if you never look at the Greeks, theta will find you. It's the quiet certainty that options decay toward zero as expiration approaches. Theta is not constant;

it accelerates as you get closer to expiration, especially for at-the-money options. This is why "cheap weekly options" are often not cheap at all. They're like buying milk with a two-day expiration date: the discount is real, but the spoilage is faster than you think.

The calendar: crashes don't RSVP. Markets don't schedule their accidents around your expiration. You can buy a perfect hedge and watch it die on the vine, only to have the selloff begin the following week. The market doesn't hate you personally. It just doesn't know you exist.

Attention span: death by a thousand paper cuts. A long-vol strategy asks you to be patient while your P&L looks wrong most of the time. That's hard. People abandon the hedge right before they need it, usually after a stretch of calm when it feels "wasteful." Then the exact event the hedge was built for arrives, and they discover what it means to be naked when volatility spikes.

Long volatility is not just a trade; it's a psychological contract.

The simplest long-vol position: buying a put (and what you're really buying)

A long put is the canonical crash bet: the right to sell the underlying at a fixed strike price. If the market collapses below that strike, your put gains value. If the market rallies or drifts sideways, your put loses value as time passes.

But what you are *really* buying with a put is a bundle of exposures:

- *Delta*: you benefit when the underlying falls.

- *Vega*: you benefit when implied volatility rises (options get more expensive).
- *Gamma*: your delta gets more negative faster as the market drops, which can make the hedge accelerate when you need it most.

That last line is why puts can feel like a miracle during a waterfall selloff. The hedge isn't just keeping pace; it's gaining power as panic builds.

The catch is that you pay for this miracle in advance. The market knows puts are valuable in a crash, so puts (especially out-of-the-money puts) often carry expensive implied volatility. This is closely related to *skew*: downside options are often priced with higher implied volatility than upside options because investors consistently demand crash protection.

So the long-put buyer is often paying two premiums:

- the time premium (theta bleed),
- the fear premium embedded in skew.

If you're going to pay that bill, make sure you're buying a hedge you actually need, not a mood ring for your bearishness.

Straddles: buying movement without choosing a direction

If you want to bet on a *big move* but you don't want to guess up or down, the classic structure is a *long straddle*: buy a call and a put at the same strike (usually at-the-money) with the same expiration.

A straddle is pure long volatility in street clothes. You pay a large premium up front. You lose money if the underlying stays near the strike and volatility falls. You make money if the underlying moves far enough in either direction, ideally with implied volatility rising.

This is also the cleanest way to understand the long-vol hurdle: the market has already priced an *implied move*. You don't get paid for a big move; you get paid for a move that's bigger than what everyone already expected, plus the path-related effects of implied volatility.

Straddles are honest. They don't pretend to be income. They don't pretend you can time the market. They're simply a receipt for movement.

Why long vol often works exactly when you feel least capable of holding it

In a true stress event, long options can explode in value, but the market environment becomes hostile in other ways:

- bid/ask spreads widen dramatically,
- trading halts happen,
- brokers tighten margin and risk controls,
- your hedges might be profitable while your core portfolio is bleeding, creating a temptation to "take the win" on the hedge to relieve emotional pain.

That temptation is not irrational. When your portfolio is down 15% and your puts are up 120%, your brain wants relief. It wants to sell the hedge to stop the stress. But the hedging logic is often the opposite: the hedge is doing its job precisely because the environment is worsening. Selling it early is like turning off the smoke alarm because it's loud.

Long-vol traders don't get paid for being smart. They get paid for holding something uncomfortable long enough for it to become essential.

The ``good deal" fallacy: cheap options can be expensive, expensive options can be cheap

People love the idea of finding underpriced options like hidden bargains on a clearance rack. The problem is that option "cheapness" is slippery.

An option can look cheap because:

- implied volatility is low,
- the option is far out-of-the-money,
- the expiration is near,
- the market feels calm.

But cheap options often have a brutal probability distribution: they expire worthless most of the time. That's not a flaw; it's the whole design. A far out-of-the-money put is like an earthquake policy in a region that rarely sees earthquakes. The annual premium is small until the year it matters. If you buy one every month, you are signing up for frequent losses with occasional huge gains.

That can be a reasonable strategy if you size it correctly. It becomes a disaster if you size it like a trade you expect to win often.

On the other side, "expensive" options (high implied volatility) can still be a bargain if the world is more unstable than the market is pricing. In real panics, implied volatility can be high and still underestimate future realized volatility. Fear can be expensive and still underpriced.

Long vol is not about buying cheap options. It's about buying *mispriced* options relative to the movement you expect, and structuring the position so that the negative carry is survivable.

Sizing: the only way long vol can be held long enough to work

If selling premium is a survival game because losses are rare but large, buying premium is a survival game because losses are common but small.

A long-vol position should be sized like a recurring expense, not like a bet. Think of it as a hedge budget: a small, tolerable bleed that you can maintain through boredom, rallies, and the temptation to stop "wasting money."

A practical sizing frame:

- Decide what you're willing to lose per month or per quarter on crash protection (a fixed percentage of the portfolio).
- Structure trades so the maximum loss is close to that budget (long options help because the loss is typically limited to premium paid).
- Accept that most cycles will lose money—and that the purpose is to change your portfolio's behavior during the cycles that matter.

The best long-vol trade is the one you can still hold after six months of being annoyed.

Structure matters: turning raw convexity into something you can actually afford

If you simply buy puts every month, you can absolutely get the crash payoff. You can also absolutely go broke from the carry. The craft is in adjusting the shape of the hedge so it fits your budget and your objective.

Here are common ways traders do that, without pretending there's a free lunch:

Put spreads: cheaper convexity with a ceiling. Instead of buying a put, you buy a put and sell a farther out-of-the-money put. That reduces the cost (less negative carry), but it caps the payout in a deep crash. Put spreads are like buying insurance with a deductible and a policy limit. They're often sensible when your goal is to protect a drawdown range (say, the first 10–20% down) rather than profit from an apocalypse.

Long-dated options: slower bleed, slower reaction. Buying more time reduces theta per day, which makes the position easier to hold. But longer-dated options can be less sensitive to an immediate volatility shock than short-dated options. They're steadier, not necessarily spicier. Think of them as sturdier insurance rather than a lottery ticket.

Staggered expirations: avoiding the "my hedge expired yesterday" problem. Instead of one expiration date, you hold multiple. This costs money, but it reduces the calendar risk of being perfectly right one week late.

VIX calls as crash convexity (with different mechanics). Many traders reach for VIX calls because they can explode in a volatility spike. But they are not magical. VIX is a volatility index, not a stock. VIX derivatives have their own mechanics, including term structure effects, settlement conventions, and behavior that can surprise traders who think "VIX up = my call up." VIX calls can be effective crash convexity, but they are not a shortcut around understanding what you own.

The through-line: you can choose where you want to bleed, where you want to convex, and what kind of tail you want. The market will still charge you. Your job is to pick a bill you can pay.

The most common long-vol mistake: buying protection when it feels safe

People tend to buy crash protection when it feels *safe* to buy it—after a dip, when fear is already in the headlines, when the VIX is elevated, when everyone is talking about tail risk. That's often when options are the most expensive, skew is steep, and you're paying up for a hedge that may already be partially "used."

Then, after the market calms down and the VIX drifts lower, people stop buying protection because it feels unnecessary. Ironically, that calm is often when protection is cheaper and easier to hold.

This isn't a moral failing; it's human. We buy umbrellas after we get wet.

A disciplined long-vol approach usually has to be systematic enough to override that instinct. Not rigid, not blind—but consistent. If you only buy protection when you're scared, you're letting the market set the price at the worst moment.

What ``catching the crash" actually looks like in P&L

The Hollywood version of a crash trade is a single heroic purchase: buy puts, market collapses, you retire.

The reality is more like dentistry. You pay for cleanings for years. Most visits feel pointless. Then one day you avoid a root canal.

A well-structured long-vol sleeve can do several useful things even if you never hit the once-a-decade crash:

- reduce portfolio drawdowns during sharp selloffs,
- provide liquidity when everything else is down (because profitable options can be sold for cash),
- let you take risk elsewhere with more confidence (because you have a seatbelt).

But the big payout events tend to be rare and clustered around market stress. That means your long-vol performance will often feel like "nothing, nothing, nothing, suddenly everything." The psychological mistake is judging the strategy during the "nothing" period.

Risk disclosures in plain English: how long options can still hurt you

Long options have limited loss (premium paid), which makes them feel safe. They are safer than naked short options in the sense that they won't create an unbounded liability. But they can still do damage:

- You can lose 100% of the premium, repeatedly.
- You can be directionally right and still lose due to time decay or implied volatility falling.
- You can buy illiquid options and discover that the theoretical price is a fantasy when the spread is wide.
- You can build a hedge that works in a slow grind down, and then watch it underperform in a sharp gap because you chose the wrong strikes/expirations for the move you got.

Long vol isn't dangerous because it can wipe you out in one day. It's dangerous because it can convince you to keep paying for something you don't understand, and then abandon it right before it would have mattered.

Actionable takeaway

Pick a long-vol structure you can hold through boredom, and convert it into a rule you can follow without guessing:

- Set a fixed "insurance budget" (a small percent of the portfolio per month or quarter) that you are willing to lose on long volatility.

- Use that budget to buy defined-risk convexity (long puts, put spreads, or straddles) with expirations far enough out that theta doesn't force constant decision-making.
- Write one exit rule for profits *and* one rule for rolling/renewing, so you don't rely on mood when volatility spikes or when calm makes the hedge feel pointless.

If the position is sized so the bleed is tolerable, you'll still be holding it on the day it stops bleeding. That's the entire game.

6.3 Riding the Curve: VIX Carry Strategies

If the VIX were a stock, volatility trading would be simple: buy low, sell high, brag on the internet. The VIX is not a stock. It's a number extracted from option prices—a synthetic "temperature reading" of expected volatility in the S&P 500 over roughly the next month. You can't own a temperature reading. You can only trade contracts *linked* to it, and those contracts come with an extra dimension most traders ignore until it bites: *time*.

That extra dimension is the VIX futures curve. It's not a trivia detail. It's the whole game.

When people say things like "VIX products always go down," they're usually describing the curve doing its quiet, daily work. When they say "VIX exploded and my product barely moved," they're usually discovering that they bought the wrong part of the curve. And when they say "I shorted VIX and it worked until it didn't," they're meeting the moment the curve flips from a gentle escalator into a trapdoor.

The simplest way to think about VIX carry strategies is this: you're not trading fear. You're trading *the price of fear at different dates*.

Spot VIX vs. VIX futures: the handoff problem

Spot VIX is the headline number. It's what flashes on TV when the market is stressed and commentators start using words like "capitulation." But most tradable products don't hold spot VIX. They hold *VIX futures*—contracts that settle to VIX on a specific date.

A VIX future is a bet on where the VIX will be at settlement, not where it is *right now*. That subtlety is why VIX can spike on a scary day and your long VIX futures position might not spike nearly as much. The market may believe the panic is real but temporary. Futures price a *future* expectation of volatility, and expectations can be calmer than the present.

So the first mental upgrade is to stop thinking "VIX up or down" and start thinking "which maturity am I exposed to?" The second upgrade is realizing that even if spot VIX does nothing, your futures exposure can still make or lose money simply because time is passing and you are rolling from one contract into the next.

That roll is carry.

The term structure: your P&L is sitting on a curve

The VIX term structure is just a fancy name for a simple picture: plot VIX futures prices by expiration date. Near-month on the left, farther months on the right. Draw a line through them.

That line has a shape, and the shape tells you whether volatility is being priced as a near-term scare or a longer-term problem.

Here's the key visual. Burn it into memory before you trade a single VIX product.

Two shapes. Two regimes. Two very different kinds of "easy money."

Contango: the calm-market conveyor belt

When the curve slopes upward (contango), near-term VIX futures are cheaper than longer-dated futures. This is the normal state of the world: markets expect today's volatility to mean-revert upward a bit into the future, or more realistically, they price a buffer for uncertainty further out.

In contango, something important happens even if nothing dramatic happens in spot VIX. As time passes, a given futures contract gets closer to expiration, and its price tends to drift toward spot VIX. If the curve is upward sloping, that drift is often *down*. That's roll-down.

Roll-down is not a prediction that VIX will fall. It's a mechanical consequence of being long a contract that is priced above the nearer part of the curve. If you're long the front-month futures in contango, you can lose money day after day even if the world is quiet and VIX doesn't do anything exciting. Your position is decaying because you're holding something that, by design, is converging to a lower number.

That decay is the engine behind the long-term grind lower in many long VIX ETPs. It's not magic. It's not a conspiracy. It's the curve charging rent.

Backwardation: the panic-market springboard

When the curve slopes downward (backwardation), near-term futures are more expensive than longer-dated futures. This usually shows up during stress, when the market believes the next few weeks are uniquely dangerous. In backwardation, time passing can actually help long volatility exposure: the roll can become a tailwind rather than a headwind.

This is where traders get the wrong lesson. They see that long VIX exposure can make money spectacularly in a spike, and they conclude the product is "broken" the rest of the time. It's not broken. It's behaving like a long insurance position that becomes valuable during a fire and depreciates when the building doesn't burn.

Backwardation is a regime where the insurance is being used. Contango is a regime where the insurance premium is being collected by someone else.

Carry: the P&L you earn (or pay) for simply existing

Carry in VIX futures is the quiet drift caused by the term structure and the rolling process. If you're long and the curve is in contango, carry is typically negative. If you're short and the curve is in contango, carry can be positive. That's why short VIX strategies have been so tempting historically: in calm markets, they're often positioned with the wind at their back.

But carry is not a free lunch. Carry is what you earn for holding an exposure that tends to explode when the regime flips. Shorting VIX in contango is like collecting rent in a neighborhood that occasionally floods. Your monthly checks feel steady. The flood, when it comes, doesn't care about your spreadsheet.

A useful way to stay honest is to separate your return into two stories:

- *Spot move story*: did volatility itself rise or fall?
- *Roll story*: did the curve shape help or hurt me as time passed?

Most retail traders only watch the spot move story. Professionals obsess over the roll story because it's the part you can see coming.

How VIX ETPs turn the curve into a tax bill

Many traders meet VIX carry strategies through exchange-traded products. The appeal is obvious: click a button, get "VIX exposure." The problem is that these products typically track a rolling position in short-dated VIX futures, not spot VIX. That rolling position makes the term structure matter every day.

In contango, a long VIX futures roll tends to sell cheaper contracts and buy more expensive ones as it maintains exposure to the front end of the curve. That's a structural drag: you're repeatedly paying up to stay exposed. Over time, that can be a powerful headwind that has nothing to do with whether the market is "really risky."

This is the part people underestimate: you can be correct that volatility will be "higher someday," and still lose money for months because you're paying the contango tax while you wait. The curve can bankrupt your patience long before the crisis arrives.

Short or inverse products flip the logic: contango can become their friend, backwardation their enemy. That is exactly why they can look stable right up until they don't. In a volatility shock, not only can futures prices jump, the curve can invert, turning what used to be a tailwind into a headwind while volatility is exploding. That's the nightmare scenario: you're short, volatility is rising, and carry flips against you at the same time.

If you ever wondered how a product can feel like it "went to zero overnight," this is one of the mechanisms that makes it possible.

The settlement and expiration details nobody reads until Wednesday morning

VIX futures have a personality. Part of it is the curve. Part of it is the settlement mechanics.

VIX futures settle to a special value derived from S&P 500 index options prices at the open on settlement day. That means settlement can gap. The VIX you see on your screen during the day is not necessarily the exact value your futures contract will settle to. If you hold into expiration, you're not just betting on a number—you're betting on a particular calculation at a particular moment.

This matters most when traders treat VIX futures like a normal equity future and assume they can "just close it on expiration day." The timing is different, the liquidity can behave differently, and the relationship between spot VIX and the settlement value can do strange things around the open.

The market is not hiding the rules. It's just daring you not to learn them.

A regime map that actually helps: when to harvest, when to hide

VIX carry strategies live and die by regime. Not prediction—regime.

In persistent contango:

Carry tends to reward short volatility exposure and punish long exposure. That doesn't mean you should blindly short. It means the default drift is often against long products, so a long-vol position needs a catalyst or a reason to exist beyond "VIX is low."

In emerging stress (curve flattening):

The easy carry can disappear before the big volatility spike arrives. This is the danger zone for carry harvesters: you're still positioned like calm will last, but the curve is starting to price that calm as fragile.

In backwardation:

The regime has changed. Long volatility exposure can benefit from both spot movement and roll. Short volatility exposure can get hit from both directions. This is where "I'm a premium seller" becomes a dangerous identity rather than a strategy.

This is not a call to be heroic or perfectly timed. It's a call to let the curve tell you when the market's climate has shifted.

The seductive trade and the grown-up version of it

The seductive version of VIX carry is simple: "Short VIX products because they decay."

The grown-up version is less romantic and much more survivable: "Express a view on the curve with defined risk, and treat backwardation like a fire alarm."

You can implement carry views in many ways (futures, options on VIX futures, spreads). The common thread is risk containment. Carry strategies can produce steady returns, but the left-tail risk is not theoretical. Volatility can spike fast enough that "I'll get out if it goes against me" becomes a fantasy. Your exits are competing with everyone else's exits.

The correct ambition for VIX carry is not to squeeze every basis point from contango. It's to avoid being the person who funds everyone else's convexity when the regime flips.

Actionable takeaway

Before touching any VIX-linked product, do one simple check and make it a habit rather than a hunch: look at the VIX futures curve and label it *contango*, *flat*, or *backwardation*. Then set a rule that connects that label to your risk.

A practical rule that keeps traders out of trouble is:

- Only run carry-harvesting (short vol / short VIX futures exposure) when the curve is clearly in contango.
- Reduce or exit that exposure when the curve flattens meaningfully.
- Treat backwardation as a regime where the market is paying you to *own* convexity, not sell it.

You don't need a PhD to trade VIX carry. You need to respect the curve like it's a weather forecast that's occasionally wrong but never irrelevant.

6.4 Time Games: Calendars and Diagonals

There's a particular kind of market paranoia that doesn't last long. It's the week-before-earnings vibe. The night-before-a-CPI-print vibe. The "everyone suddenly remembers the Fed exists" vibe. You can feel it in the options chain: the front month gets expensive, the near-dated implied volatility pops like a pufferfish, and people pay up for protection they only need for a handful of trading days.

Meanwhile, the longer-dated options often sit there looking bored. Three months out, six months out, the market is still pricing the world as mostly normal. It's not that traders believe nothing bad can happen later; it's that the *urgency* is concentrated in the near term. The fear has a timestamp.

Calendars and diagonals are how you trade that timestamp.

They're the strategies you reach for when you want to say: "I'm not sure where the stock goes, but I think the market is overpaying for *next week* relative to *next season*." It's a way to sell some volatility without becoming a full-time insurance company, and to buy some volatility without paying the full premium bleed.

Most importantly, it's how you stop treating "volatility" as one number and start treating it as a *surface*: different prices for different strikes and different dates.

The basic idea: rent out the front month, own the back month

A *calendar spread* (also called a time spread) is deceptively simple:

- Sell an option with a nearer expiration.
- Buy an option with a farther expiration.
- Usually the same strike, same type (call-call or put-put).

That structure typically creates a net *debit*—you pay money up front. That alone confuses people who associate "selling options" with receiving a credit. Calendars are different. You're not trying to collect premium and walk away. You're trying to own something durable (the longer-dated option) while subsidizing it by selling something that decays fast (the shorter-dated option).

If selling volatility is being the insurer, and buying volatility is buying the policy, a calendar spread is closer to buying the policy and then renting out part of it for a month to offset the cost.

The magic is in the decay rates. Near-dated options lose time value faster. Longer-dated options decay more slowly. In a calendar, you're betting that this difference—the *theta differential*—works in your favor.

You're also betting on something more subtle: the market's fear is misallocated across time.

Volatility has a calendar too

Implied volatility is not a single dial. It's a set of dials indexed by expiration. Markets can be terrified about next Tuesday and perfectly calm about next quarter. That isn't irrational; it's how uncertainty behaves. Some uncertainty gets resolved quickly. An earnings release is a literal information drop. A Fed meeting is a known deadline. Elections have a date circled in red ink.

Options price deadlines.

In those environments, front-month implied volatility can inflate dramatically because it contains the event. Back-month implied volatility may move too, but often less. That creates a term structure in implied vol—a front-month bulge.

Calendars and diagonals are a way to trade that bulge: sell the inflated, event-loaded near-term option and buy the calmer, longer-term option that survives after the event passes.

This isn't a guarantee of profit. It's a trade on relative pricing—and like all relative trades, it can make you feel clever right up until the underlying decides to gap 12% overnight.

How a calendar actually makes money (the version you can hold in your head)

A classic at-the-money calendar spread has a profit shape that surprises people. It doesn't love huge moves. It loves *boring*.

Here's the intuition:

- If the underlying stays near the strike into the short option's expiration, the short option decays hard (good for you), while the long option retains meaningful value (also good for you).
- If the underlying runs far away from the strike, both options can behave more like pure directional instruments, and the spread can lose.

So calendars are not "buy vol and profit from movement." They're often closer to: "sell the market's near-term panic and hope price doesn't sprint away before your short leg dies."

That's why calendars are so popular around events: the trader expects volatility to be overpriced, but also expects the actual move to be within a range. They want the event premium to collapse without the underlying detonating.

When that bet is right, calendars can be some of the cleanest, most satisfying trades in options. When it's wrong, the loss tends to come quickly and with a particular flavor: you feel like you were right about volatility being expensive and still got punished for the move.

Because you were.

The ``vol crush" trade without the bravado

People talk about "selling the vol crush" like it's a bravado move: sell a naked straddle into earnings, collect a fat premium, hope the move is small, and pray you don't get carried out. Calendars are the more civilized cousin. You can express the same core view—front-month implied volatility is too pumped—but with a built-in life raft: the long option in the back month.

That long option does two things:

- It caps some of the worst-case scenario where implied volatility rises across the curve and the underlying trends hard.
- It keeps you in the game *after* the short option expires, which matters because post-event drift can still be meaningful.

A calendar is not a free lunch. But it's a way to sell expensive short-term uncertainty while owning longer-term optionality. That mix is psychologically easier than pure short vol and financially less grindy than pure long vol.

It's also a reminder that the market's biggest inefficiencies often live in *relative* pricing, not in directional prophecy.

The Greek story: calendars are a tug-of-war

Calendars are often described as "theta positive and vega positive," which sounds like a unicorn: you earn time decay and you benefit from rising implied volatility. That's sometimes true, often approximately true, but the reality depends on strikes, moneyness, and where you are relative to the event.

Still, the tug-of-war framing is useful:

Theta differential (your paycheck).
The short option decays faster than the long option. If price behaves and time passes, that's your edge.

Vega (your sensitivity to implied volatility).
Because the long option has more time, it usually carries more vega. If implied volatility rises across maturities, the long option may gain more than the short option loses you. That can cushion nasty moments.

Gamma (the thing that ruins your sleep).
Near expiration, the short option's gamma can become sharp. If the underlying whips around, the short option's value can change fast. Calendars can look calm and then suddenly feel like you sold a live wire in the final days.

This is why many calendar traders manage the position well before the short leg hits the most explosive part of its life. The goal isn't to win a trophy for holding to the last minute. The goal is to harvest the relative mispricing without letting gamma turn your spread into a crisis.

Diagonal spreads: calendars with a personality

A *diagonal spread* is a calendar spread that also changes the strike. You still sell near-term and buy longer-term, but the strikes differ. That sounds like a small tweak. It's not. It gives the structure a personality: a built-in directional bias.

For example:

- Buy a longer-dated call at a lower strike (more expensive, more delta).
- Sell a nearer-dated call at a higher strike (collect premium, less likely to be challenged immediately).

This kind of diagonal can behave like a covered call "with benefits": you're long longer-dated upside exposure (so you don't feel naked if the stock trends), and you're repeatedly selling shorter-dated calls against it to finance the position.

Or you can construct a bearish diagonal with puts the same way.

Diagonals are calendars that admit a truth: sometimes you do have a directional view, but you don't want to pay full price for it, and you'd like to monetize the market's habit of overpricing near-term anxiety.

They're also an honest way to trade the volatility surface: you can choose where to be long volatility (back month, maybe closer to the money) and where to be short volatility (front month, maybe farther out). That's not a single trade. It's a custom exposure.

The event trade that looks easy (and why it isn't)

Calendars are frequently marketed as an "earnings strategy." The pitch is seductive: sell the pumped front-month option, buy the calmer back-month option, wait for vol crush, profit.

The problem is that earnings aren't just a volatility event. They're a *gap* event. Price can teleport. A calendar spread is not thrilled about teleportation.

Here's what can go wrong:

- The stock gaps far from your strike. Your short option may decay (good), but your long option may no longer have the right kind of exposure to recover the loss (bad).
- The entire implied vol surface can shift in a messy way. Front-month vol collapses, but back-month vol can also drop if the event resolves uncertainty more broadly.
- If you placed the calendar too close to expiration, the short leg's gamma can turn one overnight move into a position you can't "manage" elegantly.

Calendars can still be excellent around events, but they should be treated like a precision instrument, not a slot machine. The edge is not "vol always crushes." The edge is "vol sometimes crushes more than the move justifies, and the back month sometimes holds its value better than the front month collapses."

That's a mouthful, but it's also the truth.

Where calendars shine in real life: fear that expires

The cleanest calendar setups often appear when the market is pricing a near-term headline that you believe is being over-hyped, while the underlying business or macro backdrop doesn't suggest a regime shift.

Examples:

- a scheduled regulatory decision with binary headlines but limited long-term impact,

- a macro print that has been talked about for weeks (so the surprise hurdle is high),
- an earnings report for a company whose historical post-earnings moves are smaller than the current implied move.

The common feature is that the fear has a *deadline*. The calendar spread is a bet that the deadline will pass without a permanent change in the volatility regime.

It's a trade on the market's tendency to overpay for certainty in the next few days and underpay for the fact that life continues afterward.

Assignment risk: the boring footnote that can ruin your week

Calendars and diagonals introduce a practical risk that feels insultingly mundane: assignment.

If you sell an American-style equity or ETF option, you can be assigned early. That's most likely when the option is in the money and there's a reason to exercise (dividends are the classic trigger on calls). If you're running a calendar, early assignment can leave you unexpectedly long or short shares while you still hold the longer-dated option. That changes your risk, your margin, and your sleep.

Index options can reduce early-exercise risk when they are European-style, but many traders experience calendars first in single-name equities, where assignment is part of the landscape.

The takeaway isn't "avoid calendars." It's "respect the plumbing." A calendar spread is a two-leg position with a short option. Short options come with obligations, even when your maximum loss feels contained by the long leg. Your broker's margin department does not care that you intended something elegant.

Picking the strike: don't confuse ``at the money'' with ``best''

At-the-money calendars are popular because they maximize sensitivity to time decay and implied volatility changes around the strike. But diagonals show you the truth: strike selection is an expression of what you fear.

If you're worried about a sharp move up, you can choose a diagonal that sells a call strike higher, giving you more breathing room. If you're worried about a move down, you can structure the put side similarly. If you believe the stock will pin near a level (a common post-event behavior), a calendar near that level can be attractive.

The mistake is thinking calendars are a neutral, set-it-and-forget-it trade. They are a thesis about where price will spend time, not just where it will end up. Price *spent* near your strike is what pays you.

The trader's discipline: treat calendars like perishable goods

The short leg decays fast, which is the point. It's also why calendars require maintenance. A calendar spread is not something you "invest in." It's something you *operate*.

That operation has a rhythm:

- If the short option decays nicely and the underlying stays near your strike, you may take profits before expiration rather than waiting for the last scraps.
- If the underlying moves away, you may adjust by rolling strikes or rolling the short leg forward to re-center the position (if the thesis still holds).
- If the trade is clearly broken (big move, regime shift), you close it rather than turning a time-spread thesis into a stubborn directional bet.

Calendars reward traders who can admit when the market has changed its mind about the timeframe of fear. If the market decides the fear is no longer "next week" but "next quarter," the front-month edge you were selling may be gone.

Actionable takeaway

Before placing any calendar or diagonal, write the thesis in one sentence that includes a *date* and a *range*. Not "vol is high." Something like:

> "I think the market is overpricing movement *into* the front expiration, and I expect the underlying to trade roughly between $X and $Y by that date, with longer-dated volatility holding up better than front-month volatility."

Then match the structure to that sentence:

- Use a calendar when you want the underlying to *hover* near a strike into the short expiration.

- Use a diagonal when you want to sell near-term premium but you also want a directional lean or extra room on the short strike.
- Size it so the full debit is an acceptable loss, because sometimes the market will gap and your elegant thesis will get bulldozed by reality.

If you can't specify the date and the range, you're not trading time. You're just stacking legs and hoping the math turns into money.

6.5 Gamma Scalping: Trading Realized Moves

Picture a market maker on a day when nothing *big* happens, yet the tape feels like it drank three espressos. The index whips up 0.7%, down 0.9%, back up again, all before lunch. To most investors, that's noise. To a gamma scalper, that noise is inventory to be harvested—if they can stay disciplined enough to keep doing the one thing humans are famously bad at: buying when it's uncomfortable and selling when it's euphoric.

Gamma scalping is not a heroic "call the crash" trade. It's closer to running a small casino game on the microstructure of price movement. You buy convexity (usually by buying an option), and then you try to pay for it by trading the underlying against that option as price moves. When the market drops, you buy some stock. When it rises, you sell some stock. You repeat. If the market is choppy enough, the repeated "buy low, sell high" can offset the option's time decay.

It sounds like perpetual motion. It's not. The machine runs on a fuel that's easy to describe and hard to find: realized volatility that exceeds implied volatility *after costs*.

The one-sentence definition that's actually accurate

Gamma scalping is the practice of staying *long gamma* (owning an option or options) while keeping your position roughly *delta-neutral* by rebalancing with the underlying, with the goal of turning actual price fluctuations into profit.

That sentence has three loaded terms, and each one matters:

- *Gamma*: how quickly your option's delta changes as the underlying moves.
- *Delta-neutral*: your net exposure to direction is kept small, so you're not just making a disguised directional bet.
- *Rebalancing*: you are trading the underlying repeatedly. This is not a set-and-forget strategy.

If any of those pieces are missing, you're not gamma scalping. You're just long options, or you're just day trading, or you're just telling yourself a nicer story about what you're doing.

Why professionals love it and retail traders misunderstand it

Gamma scalping is often associated with market makers for a reason. Market makers live in the spread. They can trade cheaply, quickly, and often. They can hedge dozens of times a day without paying a ruinous toll in commissions, slippage, and poor fills. They also have the infrastructure (and the emotional numbness) to follow a rule without improvising every time the market twitches.

Retail traders often meet gamma scalping as an idea, not a process. The idea sounds intoxicating: "Buy an option, then just trade around it and the option becomes free." The process is less glamorous: you're paying theta every day, paying friction every time you hedge, and you need the market to deliver enough back-and-forth movement *at the right frequency* for the math to work.

The strategy is real. The edge is real. The difficulty is also very real.

A quick story: the choppy week that makes a long option feel brilliant

Suppose you buy a near-the-money straddle on a stock at $100 with 30 days to expiration because you expect a choppy month. You're not choosing a direction; you're choosing movement. The straddle gives you long gamma: when the stock moves, your delta changes quickly. If the stock drops to $97, your position becomes more delta-negative (it behaves more like short stock). If the stock rallies to $103, your position becomes more delta-positive (it behaves more like long stock).

A gamma scalper tries to neutralize that swinging delta by trading the stock:

- Stock drops to $97: your options now behave more short; you *buy* some stock to bring delta back toward zero.
- Stock rallies to $103: your options now behave more long; you *sell* some stock to bring delta back toward zero.

Notice what happened: you bought stock after it fell and sold stock after it rose. That's the entire trick, repeated with discipline. If the stock keeps oscillating, you keep harvesting small pieces of that oscillation. Those small pieces are your "scalp."

But the option position is bleeding theta the whole time. You're paying rent to own the convexity. The scalp needs to cover the rent.

The real condition for profit: realized vol must beat implied vol (plus the tolls)

People say gamma scalping profits when realized volatility exceeds implied volatility. That's directionally correct but incomplete in the way dangerous advice often is.

The real hurdle is:

> Realized movement you can capture > option theta decay + transaction costs + hedging slippage + mistakes.

The first three are math. The last one is psychology.

Theta (the rent). Every day that passes, the option loses time value, all else equal. Long gamma is not free. It is leased.

Transaction costs (the toll booth). Even if your broker offers "commission-free" trading, you still pay the spread and the market impact. If you hedge frequently in a less liquid name, the friction can be brutal.

Slippage (the hidden leak). You never hedge at the exact theoretical price. You hedge when you can get filled. In fast markets, that's often worse than you think.

Gamma scalping works best in instruments where hedging is cheap and liquid, and where the realized chop is rich. It works worst when you're paying retail spreads to chase a market that trends.

Gamma is a spring; delta hedging is the act of catching the recoil

Here's a mental model that makes gamma scalping intuitive without calculus.

Think of long gamma like holding a spring attached to the price. When price moves away from you, the spring pulls your delta in the opposite direction. If price falls, your position wants to be more short. If price rises, your position wants to be more long. That springiness is what creates convexity: your sensitivity increases as the move grows.

Delta hedging is the act of grabbing the spring and pulling yourself back to center. You buy stock when the option makes you too short. You sell stock when the option makes you too long.

If price vibrates around, you keep catching those vibrations. If price runs away in one direction, you keep paying to chase it—and your "scalps" might not cover the rent.

This is why gamma scalping is sometimes described as being paid for *mean reversion* in the small, even if you don't have a mean-reversion thesis in the large.

The unsexy secret: hedging frequency is a design choice, not a moral virtue

New gamma scalpers often hedge too much, then wonder why they're bleeding. More hedging is not always better. More hedging means more friction. Less hedging means more directional exposure. You're choosing a point on a spectrum:

- Hedge very frequently: you capture small wiggles, but friction can eat you alive.
- Hedge infrequently: you reduce friction, but your P&L becomes more path-dependent and you can drift into a directional bet.

Professionals often use rules like "hedge when delta crosses a threshold" or "hedge when price moves by X%" rather than hedging on every tick. The threshold is the steering wheel. Pick it badly and you either overtrade yourself to death or underhedge yourself into a surprise directional exposure.

The market doesn't reward effort. It rewards efficient capture.

What gamma scalping feels like in real time

A good gamma scalping day feels boring in the most disciplined way. You do the same thing over and over:

- You buy into weakness.
- You sell into strength.
- You resist the urge to "improve" the system because you read a hot take.

A bad gamma scalping day feels like the market is mocking you. You buy stock to hedge after a drop and it drops again. You sell stock to hedge after a rally and it rallies again. You start to feel like you're chasing your tail.

That feeling is normal. The hedge trade is always late by definition: you're reacting to movement, not predicting it. The strategy works not because each hedge is a winner, but because in a sufficiently choppy environment, the sequence of hedges tends to capture more back-and-forth than it loses.

The moment you demand that each individual hedge "be right," gamma scalping becomes psychologically impossible.

The biggest failure mode: mistaking a trend for chop

Gamma scalping loves oscillation. It hates clean trends.

In a strong trend, delta hedging can turn into a repeated pattern of "sell, then buy higher" or "buy, then sell lower," depending on direction. Your option might still make money if the trend is large enough (because you're long convexity), but the hedging can reduce the net profit. In some paths, it can even turn a winning long option into a mediocre outcome.

This is where traders get confused and blame the hedge. The hedge is doing what it's supposed to do: keeping you from becoming a directional gambler. The true problem is that you paid for implied volatility expecting chop, and you got a different path.

Gamma scalping is a bet on the *path* of realized volatility, not just the final distance traveled.

Picking the instrument: where retail can and can't play

Gamma scalping is most approachable where hedging is liquid and cheap:

- highly liquid index ETFs and mega-cap names (tight spreads),
- products with deep options markets (good fills, decent pricing),
- environments where you can hedge without feeling like you're paying a tax on every click.

It becomes much harder in:

- illiquid small caps (wide spreads, jumpy gaps),
- low-volume options chains (bad marks, brutal slippage),
- names with frequent gap risk (your hedge can't react to an overnight teleport).

Overnight gaps matter because delta hedging is a *continuous-time fantasy* and a *discrete-time reality*. You can hedge during market hours. You can't hedge a 7 a.m. headline that reprices the stock instantly. Your long option helps, but the scalping component can't capture what never traded.

A practical blueprint: the ``long option + delta bands" approach

A workable retail-friendly framework looks like this:

- Buy an option position that gives you meaningful gamma (often near-the-money, not too far out).
- Decide how much directional exposure you're willing to tolerate (your delta band).
- Hedge with the underlying when your net delta drifts outside the band.
- Keep a log of hedges and slippage so you can see whether you're capturing chop or just paying friction.

That's it. The sophistication comes from the calibration: which option, which expiration, which band, and whether the environment is actually delivering the movement you need.

The key is that you're not trying to be "perfectly delta-neutral." You're trying to be *neutral enough* that the P&L comes mainly from realized movement, not from your accidental directional bias.

The risk management part nobody wants to hear

Gamma scalping is not a free way to hold options. It is a strategy with its own risks:

You can bleed steadily. If realized movement is small, theta wins. Your hedges won't save you because there's nothing to harvest.

You can overtrade. If you hedge too frequently, friction becomes your dominant P&L driver.

You can underhedge. If you hedge too infrequently, you can drift into a directional position you did not intend to hold, especially during sustained trends.

You can break your own rules. The most common blow-up in gamma scalping is psychological: you stop hedging because you "feel" the next move. You turn a systematic process into a prediction contest, usually at the worst time.

A gamma scalper needs a rule set that's boring enough to follow when adrenaline is high.

The payoff profile: why this belongs in the ``advanced" bucket

The reason gamma scalping is considered advanced isn't the math. It's the operational demand. You're managing a living position:

- You must monitor delta.
- You must execute hedges.
- You must understand how your option's gamma and theta change as time passes and as price moves.
- You must account for costs honestly.

That's a lot. But it's also empowering. It turns options from a lottery ticket into an instrument you can work with. You're no longer just paying implied volatility and hoping. You're attempting to *earn* realized volatility.

It's the closest thing in retail options trading to doing what professional liquidity providers do for a living—with the important caveat that professionals have better tools, cheaper execution, and no need to sleep.

Actionable takeaway

If you want to try gamma scalping without turning it into a chaotic hobby, make it a small, controlled experiment with rules you can actually follow:

- Choose a highly liquid underlying and a near-the-money option with enough time that theta doesn't force frantic decisions (weeks, not days).
- Set a delta band (for example, "hedge when my net delta exceeds ± 20 shares per option equivalent") and commit to it for the life of the trade.
- Record every hedge with the price, the spread you paid, and your reason ("delta outside band"). At the end, compare the captured hedging P&L to the option's theta decay.

If the hedging P&L can't reliably cover the rent *after* costs, you didn't fail—you learned the most important lesson of gamma scalping: movement isn't enough. It has to be the right kind of movement, at a price you can afford to trade.

6.6 Event Trading: Earnings and Macro Catalysts

Markets can tolerate a lot of uncertainty. They can tolerate bad news, stale news, even confusing news. What they *can't* tolerate is a clock with a deadline and a crowd watching it. That's an event: a scheduled moment when uncertainty is forced to collapse into a number, a statement, a dot plot, a guidance range, a miss, a beat, a shrug, a surprise.

Options were built for deadlines. That's why event trading is where volatility stops being an abstract concept and turns into something you can point at on a calendar. The screen will literally show you the market's fear of *Wednesday at 2:00 p.m.* or *Thursday after the close*. And because humans have a habit of panicking in advance and rationalizing afterward, event options are often priced with a premium that feels almost superstitious.

That superstition can be sold. It can also be bought—but only if you understand what you're paying for.

Binary doesn't mean random; it means discontinuous

Earnings, CPI, FOMC decisions, jobs reports, major court rulings, merger outcomes—they share a trait that matters more than their headlines: they can create *gaps*. Price doesn't need to travel smoothly from $100 to $92. It can teleport. One print, one sentence, one leaked line, one unexpected phrase in Q&A, and the chart redraws itself.

That gap risk is why event trades feel so different from the day-to-day grind of premium selling or carry harvesting. A short-vol position can look perfectly safe at 3:59 p.m. and look like a completely different animal at 4:01 p.m. A long-vol position can look like a slow bleed and then, in one night, become the best decision you've made all year.

In event trading, the market isn't asking, "Will it move?" It's asking, "How much could it move *immediately*?"

The implied move: the market's receipt for fear

There's a simple number that turns event trading from vibes into something you can actually test: the *implied move*. It's the percentage move the options market is pricing for the event window.

A common approximation uses the at-the-money straddle:

$$\text{Implied Move (in \$)} \approx \text{ATM Call Price} + \text{ATM Put Price}$$

$$\text{Implied Move (in \%)} \approx \frac{\text{ATM Call} + \text{ATM Put}}{\text{Stock Price}}$$

This isn't magic; it's a translation. The straddle is what you pay to own the right to profit from movement in either direction over that expiration. If the straddle costs \$6 on a \$100 stock, the market is roughly pricing a $\pm 6\%$ move over that horizon.

The implied move is the market saying, "Here's what we think the drama is worth."

And now you can ask the only question that matters: *Is the market overpaying for drama, or underpaying?*

Right after defining it, it helps to see how this plays out across multiple quarters. The pattern is often the same: implied moves tend to be consistently larger than the actual post-earnings move, with a few quarters where reality punches through the price.

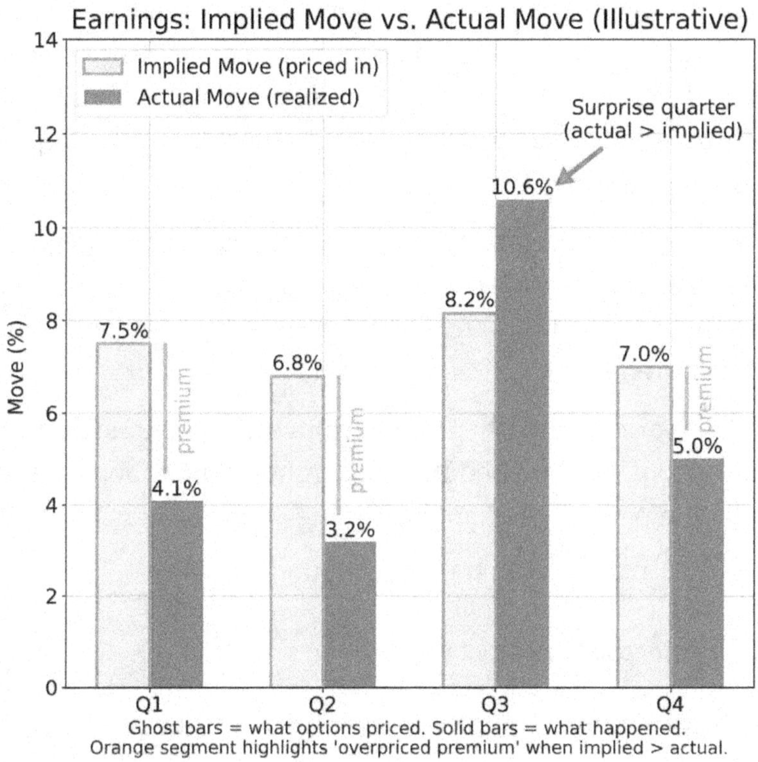

That orange "premium" segment is the temptation. It whispers the same thing every quarter: *Sell me. I'm probably too expensive.* The quarter where actual beats implied is the warning label: *I can still hurt you.*

Why implied is often higher than actual (and why it keeps happening anyway)

Event options are frequently rich for the same reason insurance is frequently profitable: the buyer is trying to eliminate a *catastrophic feeling,* not maximize expected value. A fund manager doesn't buy protection because it's a bargain; they buy it because missing a sudden move can be career risk. A retail trader buys a straddle because they don't want the humiliation of being right about "something big" and still not owning the right instrument.

There's also a structural reason: event outcomes can be lumpy. If the distribution includes even a small chance of a huge move, the option price has to reflect it. One outlier quarter can justify a lot of premium.

So the market regularly prices earnings like a haunted house: most people leave without seeing a ghost, but nobody wants to be the person who laughed at the warning sign and then got grabbed in the dark.

The two archetypes: sell the hype or buy the surprise

Most event trades fall into two psychological camps:

Sell the hype. You believe the event premium is overpriced relative to what actually happens most of the time. You want to be paid for standing on the other side of the panic.

Common structures:

- Iron condors (defined risk range-selling)
- Credit spreads on one side (defined risk directional-with-vol-sale)
- Calendars/diagonals (sell front-month event vol, own longer-dated optionality)

Buy the surprise. You believe the market is underpricing the chance of a large move, or you think the distribution is skewed in a way the options aren't capturing well. You're paying the premium because you expect a discontinuity.

Common structures:

- Long straddles/strangles (movement in either direction)
- Directional puts/calls (you have a view plus you want convexity)
- Put spreads or call spreads (reduce premium while keeping convexity)

Neither camp is "right." They're just different businesses. Selling the hype is a premium-harvesting business with event-specific crash risk. Buying the surprise is a convexity business with predictable negative carry.

The skill is choosing which business is being compensated on *this* particular event, on *this* particular underlying, at *this* particular price.

A practical checklist: pricing, path, and plumbing

Event traders lose money for three reasons that have nothing to do with intelligence. They either misread the price, misunderstand the path, or get blindsided by the plumbing.

Pricing: is the premium rich or cheap relative to history? Compare today's implied move to a simple history of actual post-event moves (say the last 8–12 quarters for earnings). Not because history predicts the next print, but because it anchors your intuition. If the market is pricing a 9% move and the stock typically moves 3–5%, that doesn't guarantee an overpricing, but it raises the bar for buying.

Also look at what changed. Is this quarter different? New product cycle, regulatory risk, merger rumors, macro sensitivity, management turnover? Event pricing isn't just statistics; it's narrative, and sometimes the narrative is warranted.

Path: what hurts your structure even if you're "right"? A short-vol earnings trade can be "right" that the move is modest and still lose because the stock drifts toward one short strike into the close, inflating gamma risk and making the position hard to manage. A long-vol trade can be "right" that the move is big and still lose if implied volatility collapses so hard that the option doesn't reprice as much as you need.

The path is the difference between a good thesis and a good outcome.

Plumbing: can you trade and survive the mechanics? Events distort liquidity. Spreads widen. Orders slip. Option markets can go from friendly to hostile in minutes. Early assignment risk increases as expiration approaches, especially for American-style options and dividend-paying stocks. Macro announcements can create one-way air pockets where your stop-loss is a wish, not a plan.

If your strategy requires perfect fills or last-second adjustments, it's not a strategy; it's a hope.

Earnings: the classic arena (and the common traps)

Earnings are the cleanest training ground because they're frequent, well-defined, and obsessively traded. They also have the most well-known trap: the *volatility crush*.

Typically, implied volatility rises into the announcement and then collapses immediately after, because uncertainty has been resolved. That crush is why premium sellers love earnings: you can be paid handsomely, and then the options deflate.

But the trap for sellers is that earnings moves are *gaps*. Your carefully engineered range can be jumped in a single print. A defined-risk condor can still be hit hard. If you are short a tight range on a name that occasionally does 12–15% overnight moves, you're not selling premium; you're writing catastrophe insurance and pretending it's a side hustle.

The trap for buyers is more subtle: you can be directionally correct and still lose if the implied move was too high. A stock can drop 4% and the put can still disappoint if the market priced 8%. When you buy earnings volatility, you're not buying a move; you're buying a move that exceeds what's already embedded.

The cleanest way to avoid this trap is to stop asking, "Will it move?" and start asking, "How does this compare to what's priced?"

Macro catalysts: when the whole volatility regime can shift

Macro events feel less "stock-specific" and more like weather. CPI and jobs reports can reprice rate expectations across the curve. Central bank meetings can change the market's understanding of what "normal" volatility should be. Geopolitical events can flip correlations and liquidity conditions.

The reason macro matters for volatility traders is that it can change the *regime* quickly. The VIX curve can flatten or invert. Index implied volatility can rise sharply even if a particular stock's story is unchanged. If you're running strategies that lean on carry or on the assumption that volatility mean-reverts, macro days are when the market tests that assumption like it's trying to break it in half.

Macro event trading often comes down to one uncomfortable question: *Am I being paid enough to hold risk through a number that can change the rules?*

Sometimes the answer is yes. Often, the best trade is simply not being exposed.

Setups that are actually coherent (not just popular)

Selling the hype with an iron condor (defined risk). This makes sense when:

- the implied move is rich versus history,
- the underlying is liquid (tight spreads, manageable fills),
- you choose strikes outside the implied move with room for error,
- position size assumes you can take the max loss without improvising.

The coherence comes from matching the structure to the thesis: you're explicitly saying, "I think the realized move will be smaller than what's priced." You are *not* saying, "I think nothing will happen." Something will happen. It just needs to happen within your boundaries.

Buying the surprise with a straddle (or strangle). This makes sense when:

- you believe the market is underpricing tail risk (company-specific blow-up potential, binary regulatory outcomes),
- the underlying has a history of occasional outsized moves,
- you can tolerate losing the premium frequently without revenge trading.

The coherence comes from accepting the business model: most of the time you're buying something that expires disappointed. The occasional payoff is meant to be large enough to justify the disappointments.

Calendars/diagonals to sell near-term event vol while owning longer-term optionality. This makes sense when:

- front-month implied volatility is inflated relative to back-month,
- you expect the event to resolve uncertainty rather than create a new regime,
- you want defined risk (debit paid) rather than steamroller exposure.

The coherence comes from trading *time* rather than just level: you're selling the market's urgency while owning the longer story.

Execution: the part that decides whether your edge exists

Event trading has a brutal truth: your spreadsheet edge can be real and still be eaten by execution.

A few practical rules keep event trades from becoming self-inflicted wounds:

- Use limit orders. Market orders during event windows are donations.
- Avoid building complex multi-leg positions in illiquid chains. Complexity amplifies slippage.
- Know your exit plan *before* the event. If you're short premium, decide whether you're closing pre-event, post-event, or managing through, and why.
- Respect expiration proximity. Near-expiration options have explosive gamma, and event gaps don't negotiate.

The market is never more eager to punish sloppy mechanics than when everyone is trading the same deadline.

Actionable takeaway

Take one upcoming earnings or macro event and run this simple, repeatable process before you place any trade:

1. Compute the implied move from the at-the-money straddle and write it down in both dollars and percent.

2. Pull the last four comparable events (four quarters for earnings, four releases for CPI/jobs) and record the actual one-day move after the event.
3. Choose your structure based on the mismatch:
    - If implied is consistently larger than actual and nothing fundamental has changed, consider a defined-risk premium sale (iron condor or spread) with strikes placed outside the implied move.
    - If actual sometimes overwhelms implied and the story has genuine binary risk, consider a long-vol structure (straddle/strangle or spread) sized so repeated premium losses are tolerable.
    - If front-month volatility is pumped relative to back-month, consider a calendar/diagonal to sell the deadline and own the aftermath.
4. Set a hard max loss *in dollars* and treat it like a seatbelt, not a suggestion.

Event trading rewards the rare combination of humility and precision: humility about what you can't predict, precision about what the market is pricing, and discipline about how much you're willing to pay (or risk) to find out who was wrong.

Chapter 7

Survival Skills: Execution and Risk Management

In the high-stakes arena of volatility trading, making money is the easy part-keeping it is the true challenge. While amateurs obsess over the perfect entry signal, veterans know that longevity is purely a function of defense. This chapter moves beyond the charts and into the trenches of risk management, teaching you how to size trades so you never face ruin, how to stress-test your portfolio against financial hurricanes, and how to execute trades without bleeding your edge to market makers. Think of this as your safety harness; you don't climb the mountain without one.

7.1 Sizing for Survival: Kelly and Ruin

A volatility trader's biggest enemy usually isn't being wrong. It's being *right too small and wrong too big*. Vol strategies have a cruel personality: they can drip-feed you wins for months, then demand a one-day payment that equals the last year of "income." That's why position sizing is not a footnote in vol trading; it's the engine block. If you size like a hero, the market eventually forces you to trade like a ghost.

The uncomfortable truth is that most blow-ups don't come from a single terrible idea. They come from a reasonable idea expressed at an unreasonable size. Selling "a little" premium, leaning short gamma "just for a week," or doubling a position because "vol is high so it should mean-revert" are not exotic sins. They're normal human impulses. The mathematics of survival exists to protect you from the most dangerous person in your trading stack: *you, in a good mood after a winning streak.*

The compounding game: why one bad day can erase a year

Trading is a compounding process, which means the *path* matters, not just the average. Lose 50% and you need +100% to get back to even. Lose 80% and you need +400%. At some point the required comeback becomes less "hard work" and more "physics violation."

Volatility products make this worse because the losses can cluster. When markets gap down, implied volatility tends to gap up, liquidity thins, spreads widen, and margin requirements jump—four horsemen arriving in the same car. That combination turns "I can take heat" into "I can't trade tomorrow." So when people say "risk of ruin," they rarely mean literal bankruptcy. They mean the practical version: the drawdown that forces you to stop, delever, or get liquidated at the worst possible moment.

This is why sizing is more important in vol than in many other corners of markets. A stock investor can be wrong slowly. A short-vol trader can be wrong *fast*, and fast mistakes don't allow graceful exits.

Kelly in plain English: the bet size that maximizes long-run growth

The Kelly Criterion is famous because it answers a clean question: *If you have a repeatable edge, what fraction of your bankroll should you bet to maximize long-run growth?* It's a compounding-optimal rule, not a comfort-optimal rule. It doesn't care about sleepless nights. It only cares about the geometric growth rate of wealth.

In the simplest "coin-flip with an edge" setup, you bet a fraction f of your bankroll each time:

- With probability p, you win and make $+b$ times your stake.
- With probability $q = 1 - p$, you lose your stake (a loss of 1 times your stake).

The classic Kelly fraction is:

$$f^* = \frac{bp - q}{b}.$$

When the bet is even-money ($b = 1$), this becomes the memorable version:

$$f^* = 2p - 1.$$

If you win 55% of the time at even odds, $f^* = 0.10$: bet 10% of your bankroll each round. That sounds aggressive, because it is. Kelly sizing is what you do when (1) you truly know your edge, (2) you can repeat the bet many times, (3) outcomes are independent-ish, and (4) your payouts don't hide a grenade in the tail.

Volatility trading violates at least two of those on a regular basis.

The part everyone forgets: Kelly is a *growth* optimizer, not a *ruin* optimizer

Kelly doesn't say: "This is the safest." It says: "This maximizes the expected logarithm of wealth." The logarithm is the key. It heavily penalizes big drawdowns (because recovering from them is exponentially hard), but it still tolerates discomfort if the long-run growth rate is higher.

That makes Kelly a good teacher even when you don't plan to follow it literally. It teaches three survival lessons:

1. There is such a thing as "too small," where you leave compounding on the table.
2. There is absolutely such a thing as "too big," where growth collapses.
3. The drop-off after "too big" is not gentle. It's a cliff.

That cliff is exactly what short-vol traders discover when they "only" double size. The market doesn't respond with "twice the stress." It responds with a nonlinear punch because the distribution isn't polite and the liquidity isn't guaranteed.

The Kelly curve and the cliff

To make the cliff visible, it helps to look at the Kelly curve: long-term growth rate as a function of bet size. Under an even-money game with a modest edge (say $p = 0.55$), the growth rate rises with bet size, reaches a peak at f^*, and then deteriorates fast. Past a point, the expected log-growth becomes negative—meaning you are mathematically volunteering for long-run decay.

Why vol traders should treat ``full Kelly" like a sports car in the snow

Kelly is elegant in a world where you know p and b. Trading is not that world. Your "edge" is an estimate built on historical data, assumptions about regime, execution quality, and the belief that the future will rhyme enough with the past for your model to stay relevant.

Volatility strategies amplify estimation error because the payoff distribution is often lopsided. If you harvest small premia most days and occasionally eat a large loss, your historical win rate and average profit can look stable right up until the moment they stop being stable. That doesn't mean the strategy is invalid; it means your sizing must assume you could be wrong about how wrong you can be.

Full Kelly is a knife-edge: it maximizes growth *if your inputs are correct*. If your inputs are too optimistic, full Kelly turns into overbetting, and overbetting is where the curve goes to die.

A useful mental reframe: the Kelly fraction is not "the right size." It's "the right size *in the fantasy world where you know the distribution*." In the real world, you don't know the distribution-you negotiate with it.

Fat tails: the market's way of making your spreadsheet lie

A "fat tail" means extreme outcomes happen more often than a normal (bell-curve) model predicts. Markets have fat tails because humans panic together, liquidity evaporates together, and leverage unwinds together. Volatility is basically a meter for this social behavior.

Fat tails matter for sizing because they create a hidden trap: you can backtest a strategy and never see the truly ugly outcomes in your sample, especially if your sample is short, your instruments changed, or your execution assumptions are optimistic. The absence of disaster in the data is not evidence of safety. It's often just a calendar artifact.

For a vol trader, fat tails show up as:

- loss events that are not merely "a bit bigger," but *multiples* bigger,
- correlations that spike toward one when you most need diversification,
- margin and liquidity constraints becoming binding at the same time as price moves.

Kelly assumes your bet outcomes are what you think they are. Fat tails are the market saying: *you don't get to know that with confidence.*

Ruin is usually a margin call wearing a mask

In a casino version of Kelly, ruin means you hit zero. In markets, you can "die" at 30% drawdown because your broker changes margin, your risk manager tightens limits, or you lose the emotional ability to execute.

Volatility products are notorious for making "paper losses" become "forced losses." Short options and short-vol positions can have convex losses (they accelerate), and margin systems are built to react to volatility. When implied volatility rises, the system often demands more collateral. That demand can force you to close positions when spreads are worst and prices are most dislocated.

So practical ruin is not a single number; it's a region. It's where:

- your liquidity buffer is gone,
- your ability to hold through noise is gone,
- your ability to roll or hedge is compromised,
- the market starts choosing your exits for you.

Sizing for survival means sizing so that your worst plausible day is still a day you can *trade after*.

Fractional Kelly: the grown-up version

The most useful way to deploy Kelly in trading is fractional Kelly: bet some fraction of the theoretical f^*. Half-Kelly, quarter-Kelly, even tenth-Kelly. This isn't cowardice. It's acknowledging that you are estimating inputs in a regime-switching world with transaction costs, slippage, jumps, and model error.

Fractional Kelly does two powerful things:

1. It reduces drawdowns dramatically relative to full Kelly.

2. It makes you much less sensitive to being wrong about your edge.

If full Kelly is "maximum growth under perfect knowledge," fractional Kelly is "strong growth under imperfect knowledge." Since trading is always imperfect knowledge, fractional Kelly is the version that actually shows up in professional survival stories.

A practical framework is to choose a *Kelly haircut* based on uncertainty:

- *High confidence edge, liquid instrument, defined risk*: you can justify a larger fraction.
- *Moderate confidence edge, liquidity can vanish, convex losses*: smaller fraction.
- *Anything with hidden tails or path dependency*: smaller still.

Volatility strategies live in the last two categories more often than traders like to admit.

Sizing with a drawdown budget (because your brain has a stop-loss, even if your strategy doesn't)

A drawdown budget is the maximum loss you're willing to tolerate over a short window without being forced to change behavior. Not "the maximum loss you can imagine," but the maximum loss after which you can still execute your process without flinching.

You can translate that into sizing with a simple discipline:

Size so that a nasty-but-plausible shock is painful, not career-ending.

For vol trading, "nasty-but-plausible" is not a 2-sigma day. It's something like: spot down sharply, implied vol up sharply, skew steepens, and liquidity worsens simultaneously. The exact stress scenarios depend on the instrument, but the idea is universal: your size should be set by the scenario that breaks you, not the scenario that flatters you.

A trader's cheat sheet: turning theory into a number

The most practical use of Kelly is as a *speed limit sign*. It tells you where the curve peaks and where the cliff begins. The conversion from "edge" to "size" will never be exact in trading, but you can still build a sizing policy that respects the shape of the problem.

Here's a compact way to operationalize it without pretending you know p to the third decimal place.

Principle	*Practical translation for vol trades*
Kelly punishes big drawdowns	Any position that can gap against you needs smaller sizing than your "average day" math suggests.
Overbetting kills growth fast	If doubling size feels like "just more of the same," you're probably ignoring convexity and liquidity.
Inputs are uncertain	Apply fractional Kelly (haircut the size) when the edge is estimated, regime-dependent, or execution-sensitive.
Ruin is usually forced liquidation	Size to survive margin expansion and spread widening, not just mark-to-market losses.

A vol-specific warning label: short premium is not a coin flip

Many options strategies resemble a rigged coin flip in the worst way: you "win" frequently and lose rarely, but the rare loss is large. Kelly can still be applied in principle, but only if the payoff distribution is correctly modeled—including the tail. If your model underestimates the left tail even slightly, the Kelly recommendation can leap from "aggressive but sane" to "quietly suicidal."

This is why traders who sell options for income often experience a strange emotional trap: the strategy works so often that it feels safer as time passes. The risk feels smaller precisely when it may be accumulating. If that describes your strategy, treat fractional Kelly not as optional prudence, but as the core feature.

Actionable takeaway: pick a fraction, then enforce it like a rule of physics

Choose a default fractional Kelly for your volatility trades *before* you're in a live market argument with yourself. A clean starting point that fits most discretionary vol trading is:

> *Use quarter-Kelly as a default cap, and only go above it when the risk is defined and liquidity is excellent.*

Then do one concrete check on every new position: write down a single stress scenario you consider nasty-but-plausible, estimate the P&L hit (include slippage/spreads if you'd have to exit), and confirm that the loss fits inside your drawdown budget *without* relying on "I'll just roll it" as the plan. If it doesn't fit, the solution is boring and powerful: cut the size until it does.

7.2 Stress Testing: Predicting the Pain

The market almost never kills you with what it *usually* does. It kills you with what it does when everyone wants the same exit at the same time. Stress testing is how you rehearse that stampede while your pulse is still normal.

Most traders think they're stress testing when they ask, "What if I'm down 2%?" That's not a stress test. That's a bad Tuesday. A real stress test asks uglier questions: *What if the S&P drops hard, implied volatility jumps, puts reprice faster than calls, and my broker quietly decides I'm now a higher-risk customer?* It's less like weather forecasting and more like checking whether your building has fire exits.

A good stress test does something that backtests often refuse to do: it removes your ability to negotiate with reality. In a spreadsheet, you can always assume you traded at the mid, hedged instantly, and rolled cleanly. In a drawdown, you're lucky if you get fills at all. The point of stress testing isn't to predict the future perfectly—it's to find out which positions turn you into a forced seller *before* the market drafts you into that role.

The signature crash combo: spot down, vol up

Equities have an awkward habit: when they fall, volatility rises. That negative relationship is one of the few "laws" markets respect often enough to build rules around. It's also why volatility trading can feel like you're holding an umbrella that turns into a sail when the wind hits.

A classic scenario worth rehearsing is simple and savage:

> *Spot down 10%, VIX doubles.*

Not because those exact numbers are sacred, but because they capture the regime shift: a fast drop in the index, followed by a violent repricing of insurance.

Here's the part that surprises people: for many option books, the damage isn't only the delta (directional) loss. It's the *shape* change of the option surface. When fear hits, implied volatility doesn't rise evenly. Out-of-the-money puts often get dramatically more expensive relative to calls-what traders call a *skew shock*. If you're short those puts-directly or synthetically-your losses can jump from "linear" to "convex" in a heartbeat.

So "spot down / vol up" is not one stress. It's a bundle:

- the index drops,
- implied vol level rises,
- skew steepens (puts inflate),
- correlations tighten (diversifiers fail),
- liquidity worsens (spreads widen, depth vanishes).

If you only model the first bullet, you're stress testing with training wheels.

Stress tests are not forecasts. They are *limits*.

Forecasts are arguments about what will happen. Stress tests are decisions about what you will *survive*. That difference matters because forecasts tempt you into debate. Survival math doesn't care who won the debate.

A trader can be completely right about the "most likely" path and still get wiped out by a low-probability path that was never rehearsed. Volatility trading, especially anything short premium, makes this painfully common. You don't lose because the world ends. You lose because the world gets weird faster than your book can adjust.

A practical way to think about stress tests:

> *A stress test is a pre-commitment device. It's your plan for the version of you who panics.*

When the screen goes red, your brain becomes a lawyer: it will find reasons to delay, reinterpret, and "give it one more day." Stress testing is how you put a contract in front of that lawyer, signed in calmer times.

Start with the position, not the scenario

A common mistake is to pick a dramatic headline scenario and then apply it to everything. That's cinematic, but not surgical. Better: start with your position and ask what combination of shocks it secretly hates.

For volatility trades, the stress-test "axes" that actually move P&L tend to be:

- *Underlying move* (spot),
- *Implied vol level move* (vega shock),
- *Skew move* (relative repricing of puts vs calls),
- *Time* (theta and the speed at which gamma bites),
- *Execution quality* (spread widening, partial fills),
- *Margin regime* (how much collateral is demanded mid-crisis).

You don't need a PhD model to begin. You need honest inputs and a refusal to hide behind mid marks.

The fastest "first-pass" stress test: shock-and-reprice

The cleanest first-pass method is shock-and-reprice:

1. Choose a scenario (e.g., SPX -10%, VIX $\times 2$).

2. Translate that scenario into your instruments' prices.
3. Recompute the portfolio value using those shocked prices.

For plain equities, step 2 is trivial. For options, step 2 is the whole game.

If you have an options model (even a basic one), you can approximate the repricing by shocking:

- spot,
- implied volatility at each strike (not just a single vol number),
- and possibly interest/dividend assumptions if relevant (usually secondary in a crash scenario).

If you *don't* have a model, you can still do something better than hand-waving:

- Look at how your option's implied vol typically moves when spot drops (even rough heuristics help).
- Apply a bigger vol shock to puts than calls to reflect skew steepening.
- Reprice using broker-provided theoretical values *at bid* if you'd be selling in the stress.

That last line is where realism begins. In stress, the mid is a mirage. Your liquidation value is closer to bid for longs and ask for shorts. If you stress test at mid, you're measuring a portfolio you don't actually own.

A more intuitive approach: translate everything into "pain per unit"

Even if you don't compute exact prices, you can make stress testing concrete by converting positions into a common language: how much you lose per unit move.

For options and vol books, the useful "pain units" are:

- *Delta*: dollars lost if spot moves against you by 1%.
- *Vega*: dollars lost if implied vol rises by 1 point (or 1 vol point).
- *Gamma*: how quickly delta gets worse as spot keeps moving.

The trap is thinking these are stable. They aren't. In a selloff, gamma can explode as options move toward the money and time to expiry shrinks. Vega exposure can become more expensive because the vol surface reshapes. Delta can jump because your "neutral" hedge was neutral only at yesterday's price.

So the stress test should not treat Greeks like constants. It should treat them like *animals*: calm when fed, dangerous when cornered.

A simple way to handle this without heavy math is *two-step Greeks*:

1. Compute Greeks today.
2. Shock spot and vol, then recompute Greeks at the shocked state and estimate the incremental damage if the move continues.

This catches the nonlinear part: your exposures morphing as the market moves.

Skew shock: when the market overpays for fear (and you might be short the receipt)

Skew is the market's pricing of downside fear: out-of-the-money puts usually trade at higher implied volatility than calls. In calm times, skew is a tax. In panic, skew can become a *tariff*.

A skew shock is not "vol is up." It's "downside vol is up more." Practically, that means:

- Puts you're short get marked much higher (worse for you).
- Puts you're long can gain value even if spot hasn't moved as much as you expected.
- Put spreads can behave in unintuitive ways because the wing may reprice differently than the body.

Stress tests that ignore skew often underestimate the loss of short put structures and underestimate the protection offered by certain hedges. That's one reason traders sometimes discover—mid-crisis—that their "hedge" was more like a raincoat in a hurricane.

A practical skew-shock rule-of-thumb stress (no model required) is to apply an extra implied-vol increase to downside strikes. For example, if you shock "vol +10 points" at-the-money, you might shock "vol +15 or +20 points" for far out-of-the-money puts. The exact numbers aren't universal; the discipline is.

Liquidity stress: the hidden multiplier

Price moves are the headline. Liquidity is the fine print that becomes the bill.

In volatility markets, execution quality degrades exactly when you most want to trade:

- bid-ask spreads widen,
- quotes become small and flickery,
- you get partial fills,
- the "theoretical value" moves faster than the displayed market.

A stress test that marks at mid is like a fire drill where everyone teleports outside. It's comforting and useless.

A practical way to include liquidity without building a microstructure simulator is to add a liquidation haircut:

- For long options you might need to sell: mark them at bid under stress.
- For short options you might need to buy back: mark them at ask under stress.
- If you trade spreads, assume you cross a meaningful fraction of the spread to adjust.

If that feels pessimistic, good. Stress tests are supposed to feel pessimistic. Optimism belongs in trade selection; pessimism belongs in survival planning.

Margin stress: the part that turns "unrealized" into "mandatory"

Many traders learn margin the way people learn about black ice: by hitting it.

In a volatility spike, margin requirements can increase—sometimes quickly. The point of margin is to ensure the system can withstand losses during the time it takes to liquidate positions. When markets become more volatile and less liquid, the system asks for more collateral. That's rational from the system's perspective, but from your perspective it means you can be forced to reduce risk precisely when risk reduction is most expensive.

So a stress test for a vol book should include:

- your P&L hit under the scenario,
- plus a collateral demand shock (an assumed percentage increase in required margin),
- plus the cost of meeting that demand (selling something, hedging something, or wiring cash).

The psychological trap is assuming you can "ride it out" because you're solvent. Solvent traders still get liquidated if they can't meet a margin call on time. Survival is partly about having cash when cash becomes fashionable.

Building a heatmap: one picture that tells the truth fast

Once you have a basic repricing method (even if approximate), the most useful visual is a grid of scenarios: spot moves across columns, VIX (or vol level) moves across rows, and each cell shows P&L.

The power of the grid is that it reveals asymmetry. A healthy book often shows:

- manageable losses in mild stress,
- losses that grow in a way you expected,
- and no single cell that looks like a trapdoor.

An unhealthy book has a "corner" where the P&L collapses—usually the corner that corresponds to -20% spot and big vol up. That corner is where forced selling happens, and forced selling is where strategies go to die.

Even without a formal diagram, you can implement the idea as a table you update weekly:

- Spot: $-5\%, -10\%, -20\%$
- VIX/vol: $+5, +10, +20$ points (or a relative shock like $\times 1.5, \times 2$)

Then you compute P&L with your liquidation haircuts and margin add-on assumptions. You will learn more from that grid than from most performance reports.

A worked example (conceptual, but concrete)

Imagine a portfolio that looks "sensible" on a quiet day:

- short some put premium for carry,
- long some calls "for upside participation,"
- delta-hedged "roughly,"
- and a small VIX call position as a hedge.

In a mild down move (−5%, vol +5), the short puts lose, but maybe the hedge helps and you feel fine. In the severe scenario (−10%, VIX doubles), three things can happen at once:

- the short puts go from out-of-the-money to near-the-money, increasing gamma and making hedging more expensive,
- implied vol on those strikes jumps and skew steepens, inflating the puts beyond what "vol +X" at-the-money would suggest,
- the VIX calls hedge helps, but you can't fully monetize it because spreads widen and you hesitate (or you need to sell it to meet margin, which is the financial version of eating your emergency food because your landlord raised the rent).

The stress test's job is to force you to quantify that tangled mess into a number you can act on. If the number says you're one bad morning away from a margin call, you don't have a "risk-managed carry strategy." You have a countdown timer.

Governance for a solo trader: turning stress tests into rules, not trivia

Big institutions talk about "governance," which sounds like a committee and a conference room. For an individual trader, governance is simpler: a written rule you obey when you're emotional.

A stress test becomes useful when it triggers decisions:

- *Sizing limits*: "If the -10%/vol spike cell is worse than $-X\%$ of equity, I reduce exposure."
- *Hedge policy*: "If skew steepening makes my downside P&L convex, I add defined-risk protection, not just delta hedges."
- *Cash policy*: "If margin under stress exceeds available liquidity, I reduce positions until it doesn't."

Notice how unromantic those rules are. That's a compliment. You don't want your survival plan to be artistic. You want it to be executable while your hands are shaking.

The stress test that matters most: ``Can I act?"

Here's the final realism check. Under your worst plausible scenario, ask:

> *Would I be able to trade my way out, or would I be negotiating with my broker and my fills?*

If the answer is "I'd probably be forced to reduce risk," then your stress test must assume forced-reduction pricing: wide spreads, poor execution, and time pressure. Many strategies look acceptable until you assume you'll be exiting under duress. Then they reveal their true nature.

If this feels like paranoia, remember: markets reward the people who prepare for chaos during calm periods. Everyone else prepares during chaos, which is like buying insurance while the house is already on fire.

Actionable takeaway: adopt the weekly two-scenario ritual

Every week, run two scenarios on your live book and write the results down somewhere you will actually look:

1. *Nasty-but-plausible*: SPX −10%, vol up sharply, skew steeper; mark at liquidation values (bid/ask), include a margin add-on.
2. *Ugly corner*: SPX −20%, vol up more; assume liquidity is poor and you can only reduce risk by crossing wide spreads.

Then set one hard rule: if either scenario exceeds your drawdown budget or your available liquidity (cash plus realistic credit), cut risk *that day*. Not "monitor it," not "re-evaluate after CPI," not "wait for a bounce." The best time to reduce exposure is when you still have a choice.

7.3 Managing Gamma and Tail Risk

There's a special kind of confidence that comes from a strategy that wins quietly. Day after day, small green numbers stack up. Your blotter starts to look like a well-behaved lawn. Then one morning you wake up, futures are limit-down-ish, every quote is wider than you remember, and your "income strategy" is suddenly an emergency fundraiser. That's not bad luck. That's short convexity collecting its bill.

Options are not dangerous because they're complex. They're dangerous because they are *nonlinear*. A little move can become a big loss, not by changing the rules, but by revealing the rules you were always playing under. If you want to survive trading volatility, you need a gut-level feel for two forces that don't care about your intentions: *gamma* and *tails*.

Convexity: the shape of your risk

Most people learn risk as a straight line: if the market drops 1%, you lose about 1% times your exposure. That's *linear* risk. It's the mental model behind "I can handle a 5% drawdown."

Options introduce *convexity*: the relationship between the market move and your P&L curves. If you are *long* convexity (typically long options), losses tend to slow down and gains can accelerate. If you are *short* convexity (typically short options), the opposite happens: losses accelerate as the market keeps moving against you.

"Short convexity" is a clean phrase. The lived experience is messier: you start with a manageable loss, you hedge, the hedge slips, your delta changes, you hedge again, and suddenly you're trading in a storm you helped create. This is why short-vol blowups feel like they happen in a single day. The position isn't only losing—it's *becoming a worse position as it loses*.

Gamma in plain English: how fast your risk is changing

Delta is the "slope" of an option's price with respect to the underlying. If delta is how much you lose for a small move, gamma is how quickly that delta changes as the market moves.

A practical definition that sticks:

> *Gamma is the acceleration pedal of your P&L.*

If you're short gamma, your delta gets more and more painful as the market runs away from you. If you're long gamma, your delta tends to swing in your favor; the position starts helping you hedge rather than demanding constant repairs.

Two details matter more than the textbook:

- Gamma is most intense for *at-the-money* options.
- Gamma becomes more intense as *expiration approaches*.

That combination creates a classic trap: the trade looks safe because "there are only a few days left." In reality, those last few days are when a short option position can turn feral. The premium is small, the gamma is large, and the market does not need to move much to put you in the danger zone.

Negative convexity: the ``steamroller" feeling has a formula

Short option positions often have negative convexity. The mental picture is familiar: you collect small premium (pennies), but you're exposed to a large adverse move (steamroller). That's not a moral lesson; it's geometry.

A linear position loses like a straight line. A short-gamma position loses like a curve that steepens-slow at first, then faster, then frighteningly fast. That curvature is why people get blindsided. Humans are bad at intuiting acceleration, especially when the early part of the curve feels calm.

The real-world mechanism: hedging short gamma forces you to buy high and sell low

If you're short gamma and you try to delta-hedge, you tend to do something psychologically nauseating:

- when the market drops, your short options become more negative delta, so you sell futures/stock to hedge,
- when the market rebounds, you buy back to reduce the hedge.

That's selling after weakness and buying after strength-mechanically. It's not that you're undisciplined; it's that the position structure *requires* it.

In a calm market, this can feel manageable because the moves are small and mean-reverting. In a trending selloff, it becomes a treadmill that speeds up. If liquidity is good, you can survive with costs. If liquidity is bad, the same process becomes a loss amplifier: every hedge adjustment is executed wider, with more slippage, at the worst moments.

This is where theory meets the most practical advice in volatility trading:

> *If your strategy requires frequent trading to avoid disaster, it is also a strategy that can be killed by transaction costs and illiquidity.*

Gap risk: the market's way of jumping over your stop

Stop-losses are comforting because they make risk feel like a choice. Options remind you that sometimes risk is an event.

Gap risk is the risk that the market moves discontinuously—overnight, on a headline, on a central-bank surprise—so your exit is not where you planned it. If you're short options, the move can reprice the entire surface before you touch your keyboard. The loss isn't earned tick-by-tick; it's delivered.

Volatility traders feel gap risk most acutely when they're short near-dated options. The option's value can double or triple on a fast move because:

- spot moved,
- implied vol rose,
- and the option's gamma is highest near expiration.

That triple effect is why "I'll hedge if it moves" is not a complete plan. Sometimes it moves *to a new universe* first.

Tail risk: it's not the worst case, it's the *wrong time*

Tail risk is the risk of extreme outcomes-the far left tail of returns. People often imagine tails as rare disasters. In practice, what makes tails lethal is timing and correlation. Tail events tend to happen when:

- you're levered because things have been calm,
- liquidity is thin because market makers are cautious,
- multiple positions you thought were diversified suddenly move together.

Tails are when the market stops acting like a collection of independent bets and starts acting like a single crowded trade. Even if your trades are different on paper, they may share the same vulnerability: short convexity to the same underlying fear.

Tail hedges: insurance that's expensive for a reason

Tail hedges-like SPX puts or VIX calls-often have negative carry. You pay premium and, most of the time, you watch it decay. This is where many traders get cute. They search for "cheap hedges," which often means "hedges that won't pay when I need them."

The market tends to overprice crash insurance relative to calm outcomes because crashes are uniquely painful to investors. That overpricing is a premium you pay for convexity. Complaining about it is like complaining that life jackets aren't free.

A better question than "Is the hedge cheap?" is:

> *Does the hedge pay in the exact scenario that breaks my book, and can I hold it long enough for it to matter?*

If your core risk is a fast crash plus a volatility spike, a hedge that only helps in a slow grind lower is a false friend. If your core risk is a volatility spike without a huge spot move, a hedge that needs spot to collapse might not save you.

The VIX option twist: settlement can surprise you

VIX options feel like the perfect tail hedge because they can explode upward when fear rises. But they come with a mechanical quirk that matters precisely when you're leaning on them: they settle to a special opening quotation (often referred to as VRO), not necessarily the prior close or the intraday print you've been staring at.

This introduces a specific kind of basis risk: the value you think you have, based on displayed VIX, can differ from the value you actually settle to. That doesn't make VIX options bad. It makes them *instruments*, not magic. Tail hedging is not only about buying convexity; it's about understanding the plumbing of how that convexity pays.

If you're relying on VIX options as disaster insurance, you want to rehearse the settlement reality:

- How do they settle?
- When do they stop trading relative to settlement?
- What happens if the surface is chaotic at the open?

Tail hedges are supposed to reduce surprises, not introduce new ones.

Defined-risk structures: the simplest way to avoid a catastrophic curve

One reason spreads exist is emotional: humans like to know the worst-case number. A defined-risk structure (like a vertical spread) caps your loss. That cap is not free-you give up some payoff-but it turns your risk into a number you can plan around.

If you're naturally drawn to premium selling, defined-risk structures can be the difference between a scary drawdown and a fatal one. It's the difference between "I don't like this" and "I can't survive this."

There's a subtle bonus, too: defined-risk positions often make decision-making cleaner. When you know the maximum loss, you can avoid the spiral of reactive hedging that turns short gamma into a job you can't leave. You still need discipline, but you aren't negotiating with infinity.

A practical way to read your own risk: where is my book short gamma?

You don't need to calculate every Greek to improve your survival odds. You need to identify where your book becomes *fragile*. Three common fragility zones:

- *Near expiration*: gamma is highest, hedging must be fast, and small moves hurt more.
- *Near the money*: options become more sensitive; deltas swing.
- *After a volatility drop*: premium selling feels "safe," sizing creeps up, and the next spike hurts more because you're bigger.

Fragility is not a moral failing. It's a property of the payoff surface you chose. The cure is not shame. The cure is structure and sizing.

When to hedge gamma vs when to cap it

There are two broad ways to manage short gamma risk:

- *Hedge it dynamically*: adjust delta as the market moves, accepting trading costs and execution risk.
- *Cap it structurally*: use spreads or buy wings so the worst case is bounded.

Dynamic hedging can work well for liquid underlyings and disciplined processes. But it has an ugly dependency: it assumes you can trade when you need to. Structural caps cost carry, but they reduce the chance that you'll be forced to trade in a liquidity vacuum.

A good rule of thumb: if the trade's survival depends on flawless execution during a panic, you don't have a risk plan—you have a wish.

The emotional tell: if you're watching the screen more, you're probably short gamma

Here's an unscientific but reliable indicator. Long gamma positions tend to make you calmer as volatility rises; they can be rebalanced from strength. Short gamma positions tend to make you stare at every tick, because every tick changes the situation. If your attention level is spiking with market movement, your risk might be accelerating even if your P&L hasn't caught up yet.

The market loves to hand out small rewards for short gamma because it wants you to increase size. Your job is to treat those rewards like they come with a warning label.

Actionable takeaway: adopt the ``convexity audit" before you place the trade

Before entering any option position that involves selling premium, do a 60-second convexity audit and write the answers in your trade notes:

1. *Where is the worst-case loss capped?* If it isn't capped, what specific hedge turns it into a number?
2. *What happens if the market gaps overnight?* Assume you cannot trade until after the gap.
3. *If implied volatility jumps and skew steepens, does the position get worse faster than your hedge gets better?*

If any answer is fuzzy, shrink the size or change the structure until it isn't. The goal isn't to eliminate risk. It's to eliminate *catastrophic* risk—the kind that turns one bad day into an ending.

7.4 Surviving Liquidity Crises

A liquidity crisis is what happens when the market stops being a place and becomes an emergency. Quotes are still on your screen, but they feel like stage props. The "price" exists, technically, the way a lifeboat exists when it's already full. You discover—too late—that your biggest risk wasn't whether you were right or wrong. It was whether you could *act*.

Volatility traders meet this reality earlier than most because volatility is the market's fear gauge and its plumbing stress test at the same time. When fear spikes, everyone wants insurance, everyone wants out of the same structures, and market makers suddenly have very good reasons to widen spreads or step back. That's not villainy; it's survival on their side of the trade. A liquidity crisis is the moment you realize other people also manage risk—and when they do, your costs become their compensation.

Liquidity isn't a feature of an instrument. It's a mood.

Most of the time, "liquid" is treated like a permanent label. SPX options are liquid. VIX futures are liquid. This ETF is liquid. Then a real shock hits and you learn the more accurate definition:

> *Liquidity is the willingness of other people to take the other side at a reasonable price.*

Willingness is emotional. It disappears when inventories are stressed, when correlations jump, when hedges get expensive, and when funding costs rise. You can trade a highly liquid product at 11:00am on a normal Wednesday and still face a liquidity crisis at 9:31am on a panic Friday. Same instrument. Different mood.

That's why the usual advice—"trade liquid stuff"—is necessary but not sufficient. You also need a plan for when the liquid stuff temporarily stops acting liquid.

The two liquids that matter: market liquidity and funding liquidity

There are two kinds of liquidity you care about, and they feed each other in ugly ways.

Market liquidity is what you feel in the spread and the depth. Can you trade size without moving the market? Can you get filled without giving away a week of expected returns?

Funding liquidity is whether you can finance your positions—margin, collateral, credit lines, cash buffers. Funding liquidity is what you feel when your broker raises requirements or your clearing firm gets nervous.

In calm markets, these are separate topics. In crises, they become a feedback loop:

> *Bad market liquidity increases losses and hedging costs, which strains funding, which forces selling, which worsens market liquidity.*

It's a liquidity spiral, and it has a specific personality: it punishes the people who need to trade *now*. If your strategy depends on "I can always adjust," a liquidity spiral is the market asking, "Always?"

What it looks like on the screen (and why your marks start lying)

The first sign of a liquidity crisis is not always a crash. It's a change in texture:

- bid-ask spreads widen and stay wide,

- quotes flicker and vanish,
- the size available at the best price shrinks,
- your fills come in pieces at worse levels than you expected.

Then comes the second sign: your internal sense of P&L becomes unreliable. Many traders mark positions at the midpoint because it feels fair. In a liquidity crisis, the midpoint can become a polite fiction. If you have to sell, you're a seller at the bid. If you have to buy back shorts, you're a buyer at the ask. That "tiny" difference becomes a meaningful fraction of your premium—especially in options where spreads can expand dramatically.

The most dangerous part is psychological. Mid marks let you believe you have more time. Liquidation reality says you have less.

A practical rule that saves careers:

> *When spreads widen, stop asking "What am I worth at mid?" and start asking "What am I worth if I have to get out?"*

Why vol products are extra sensitive: hedging demand concentrates

Volatility markets don't just reflect fear; they *concentrate* hedging behavior. In a selloff, demand for puts rises, dealers hedge their exposures, and gamma/vega flows can become one-way traffic. One-way traffic is where liquidity goes to die, because liquidity requires disagreement.

If you're short options, you may need to buy them back precisely when everyone else is buying them. If you're long options and counting on monetizing gains, you may discover the bid is thin and the market is paying you less than the theoretical value suggests. Either way, the difference between theory and tradable reality becomes the real risk.

That's why "execution" and "risk management" stop being separate topics during a crisis. Your risk is not your model Greeks. Your risk is your model Greeks *multiplied by your ability to trade.*

Margin expansion: the silent margin call that starts hours before the email

Most traders imagine margin calls as dramatic messages. In reality, margin pressure often begins quietly: requirements increase, buying power shrinks, and suddenly you have less room to breathe. Clearinghouses and brokers adjust margin because they're trying to survive extreme moves over the time it might take to liquidate positions. When volatility rises and liquidity worsens, that time gets longer and the potential adverse move gets larger—so margin goes up.

For a volatility trader, margin expansion is particularly nasty because it's correlated with your losses:

- you lose money as volatility rises (if you're short vol),
- and your required collateral increases at the same time.

That's a double hit: equity down, requirement up. Even if you have a fundamentally good long-run strategy, a margin expansion can turn it into a forced liquidation machine.

The cruelest detail is timing. Margin is often raised when the market is already stressed—when exiting costs the most. So "I'll just reduce risk if it gets bad" is not a plan unless you can reduce risk *before* it gets bad, or you have enough cash to ignore the margin increase.

The case study everyone should study: XIV (2018) as a liquidity and convexity lesson

You don't need to trade the product to learn from it. In early 2018, the inverse VIX ETN XIV collapsed after a sharp volatility spike. The details can get technical, but the lesson is plain: short-vol exposure can look stable until it hits a regime change, and product mechanics plus market constraints can turn that regime change into a cliff.

The part that matters for your survival is not the ticker symbol. It's the pattern:

- a crowded trade built on "vol won't spike too much,"
- a sudden move that forces mechanical repositioning,
- degraded liquidity when everyone tries to move together,
- rapid losses amplified by convexity and constraints.

If you trade volatility, you should assume that your own strategies can become "crowded" at the worst time—not because you personally are big, but because the *behavior* is common. Many people sell premium. Many people hedge the same way. Many people roll at the same times. Crowding is often a strategy trait, not a fund-size trait.

A crisis is when your optionality gets repossessed

Optionality is the ability to choose: to hold, to hedge, to roll, to exit. Liquidity crises are when that optionality gets repossessed by the market. You may still have choices, but they become expensive choices.

This is where "risk of ruin" becomes operational. You're not ruined when you hit zero. You're ruined when you lose the ability to make good decisions: you're trading to satisfy constraints rather than express views.

The best crisis preparation is to make your constraints looser *before* the crisis:

- less leverage,
- more cash,
- more defined-risk structures,
- fewer positions that require frequent hedging.

These sound boring. Boring is the point. A crisis is not the time to discover you built a strategy that only works when the world is polite.

The ``liquidity haircut" method: stress your exits, not your marks

Here's a practical technique that turns liquidity risk into a number you can manage. In addition to stressing prices and implied volatility, stress your exits:

- Long options: assume you can sell at bid (or worse).

- Short options: assume you can buy back at ask (or worse).
- Spreads: assume you pay a meaningful fraction of the combined spread to adjust.
- Futures: assume slippage increases and depth thins; large orders break into worse fills.

Then compare the resulting liquidation P&L to your "normal" model-based P&L. The gap between them is the price of liquidity. In calm markets, that gap is small enough to ignore. In crises, it can be the difference between "survives" and "forced seller."

A good habit is to compute a simple ratio:

$$\text{Liquidity multiplier} = \frac{\text{Stress loss with liquidation haircuts}}{\text{Stress loss at mid/model}}.$$

If the multiplier is consistently large, your strategy isn't only taking market risk. It's taking *liquidity risk*—and liquidity risk charges interest at the worst times.

Keep your ``get-out-of-jail" assets actually usable

Traders love the phrase "dry powder." In practice, many people's dry powder is damp. It's tied up in positions that also become illiquid, correlated, or margin-intensive under stress.

In a crisis, the best liquidity is:

- cash,
- unencumbered buying power,

- and hedges that can be monetized without destroying the rest of the book.

Notice what's missing: "assets that should be liquid." Lots of things are liquid until they aren't. Cash remains cash.

This is also where you should be honest about your own behavior. If your hedge is a position you *hate* holding because it bleeds carry, you may not hold it long enough for it to save you. A hedge you can't emotionally carry is not a hedge; it's a hypothetical.

Avoid the leverage trap: when small spreads hide huge exposure

Leverage is not only "borrowed money." In options and volatility products, leverage hides in payoffs. A small premium can control a large notional exposure, and a strategy that sells many small-premium options can accumulate a large short convexity position without ever feeling "levered" in the moment.

Liquidity crises punish hidden leverage because:

- hedging requirements increase (gamma demands action),
- margin requirements increase (funding demands collateral),
- and execution worsens (the market charges a crisis tax).

If you want a single anti-blowup rule that works across most volatility strategies:

> *Don't run positions that require you to be a great trader during the worst hour of the year.*

Great execution in calm markets is valuable. Great execution in a liquidity crisis is rare. Build a book that doesn't demand rare skills for basic survival.

Contingency planning that isn't fantasy

A real contingency plan answers questions that feel slightly embarrassing to ask when everything is fine:

- If spreads double, which positions do you reduce first?
- If margin requirements jump, where does the cash come from?
- If you can only trade one instrument for hedging (because everything else is a mess), which one is it?
- If the open is chaotic, do you have orders in place, or are you relying on being calm and quick?

The goal isn't to predict every crisis shape. It's to reduce your dependence on improvisation. Improvisation is great for jazz, not for margin calls.

The crisis posture: reduce decisions, increase rules

During a liquidity event, decision-making quality deteriorates. The best traders know this and simplify their world:

- fewer positions,
- clearer exposure,

- pre-defined maximum losses,
- pre-defined reduction triggers.

If your book is a complicated story, a crisis will edit it into a tragedy. Complexity makes it hard to see what's driving P&L, which makes it hard to act, which makes you late, which makes you pay more.

There's a reason professionals talk about "staying liquid." They mean financially, but they also mean cognitively: staying able to think.

Actionable takeaway: create a ``liquidity crisis protocol" in one page

Write a one-page protocol and keep it where you trade. It should include:

- *Your liquidation marking rule*: switch from mid to bid/ask marks when spreads widen beyond a threshold you define.
- *Your margin buffer rule*: maintain a minimum cash/buying-power buffer, and reduce positions immediately if the buffer is breached.
- *Your forced-reduction order*: a ranked list of positions you will cut first because they are most convex, least liquid, or most margin-hungry.

Then do the most important part: rehearse it once in a simulated stress (even manually). Liquidity crises don't reward cleverness. They reward preparedness, humility, and the boring habit of keeping enough cash to say, "No, I'm not forced to sell today."

7.5 Smart Execution: Liquidity and Costs

If risk management is how you avoid dying, execution is how you avoid bleeding out slowly. Most trading edges aren't murdered in one dramatic moment. They're quietly donated to the market a few cents at a time, like a subscription you forgot to cancel. In volatility trading those "few cents" are often the entire strategy.

Backtests are especially guilty here because they live in a frictionless universe. They assume you can buy at the mid, sell at the mid, hedge instantly, and roll without paying a toll. Real markets charge tolls. And in options, the toll booth is everywhere: bid-ask spread, slippage, market impact, and the simple fact that your order tells other people something about you.

The technical name for this gap between the paper world and the tradable world is *implementation shortfall*: the difference between the price you thought you'd get and the price you actually got, including all the little compromises along the way. You don't need to say the phrase out loud. You just need to treat it like a predator that follows you around.

The two kinds of costs: the ones you see and the ones that eat you

Commissions are explicit. You can point to them on a statement. That makes them emotionally satisfying: you can blame someone.

Execution costs are the ones you *feel* but can't always isolate. The big three:

- *Bid-ask spread*: the price of immediacy.
- *Slippage*: getting filled worse than expected because the market moves or the quote was a mirage.
- *Market impact*: your own order pushes the price against you, especially in size.

In volatility markets, spreads matter more than you expect because you often trade more frequently (hedging, rolling, adjusting strikes), and because spreads can widen exactly when you most need to trade. That's the cruelty: the market charges the highest fees when you're most desperate.

A useful way to keep yourself honest is to mentally label spreads as a *recurring expense*. If you're selling premium, spreads are like rent. If you're dynamically hedging, they're like gasoline. Either way, they're not optional.

The mid-price is not a price

The midpoint of the bid and ask feels like truth. It's not truth. It's an average of two advertisements.

In calm markets, trading near the mid is sometimes achievable. In stressed markets, mid becomes a comforting hallucination-especially in options, where displayed size can be tiny and quotes can be wide. If you mark your P&L to mid but you could only exit at the bid (for longs) or the ask (for shorts), you're living in a parallel portfolio.

This matters for two reasons:

- It makes you think you have more cushion than you do.
- It makes strategies look profitable in research that are unprofitable after realistic fills.

A simple discipline: for any trade you might need to exit under stress, periodically compute what your book is worth at *liquidation marks* (bid for longs, ask for shorts). You don't have to live there. You just have to visit often enough that you stop lying to yourself.

Why spreads are wider in options (and why that's not ``unfair")

Options are harder to make markets in than stock because the market maker is pricing a moving object. The underlying moves, implied volatility moves, skew moves, and time passes. When uncertainty rises, the cost of holding inventory rises, so the market maker widens the spread to get paid for that risk.

That's the economic reality: the spread is compensation for warehousing uncertainty.

The practical implication is non-negotiable: if your volatility strategy expects to trade frequently, you need to treat the spread as a core input, not an afterthought. A strategy that makes 1.0 unit of edge per month but pays 0.8 units in spread and slippage is not a strategy. It's a hobby with receipts.

Market orders vs limit orders: speed versus certainty

Order types sound like paperwork until you try to trade into a fast market and discover you've been driving without brakes.

A *market order* is a request for speed. It says: "Fill me now, at whatever price is available." In a stable, liquid market, that may be fine. In a volatility spike, it can be a controlled demolition of your edge. Market orders can fill across multiple prices, and the final average can be nowhere near what you thought you saw on the screen.

A *limit order* is a request for price control. It says: "Fill me only at this price or better." The catch is that limit orders don't guarantee execution. In a fast market you can become the person watching the trade happen without you.

So the tradeoff is real:

- Market orders: high execution probability, low price certainty.
- Limit orders: high price certainty, low execution probability.

Volatility markets tend to punish the lazy use of market orders because the spread is often the entire profit margin. If you cross it casually, you're paying your profits to someone else on purpose.

A good default in vol instruments: use limit orders unless you have a specific reason not to, and treat "reason not to" as a rare exception, not a daily habit.

The hidden trap in limit orders: adverse selection

Limit orders feel like free money because they let you "buy cheaper" or "sell higher." The market has a response to that: it selectively fills you when conditions are moving against you.

That's *adverse selection*: you get executed more often when the market knows (or suspects) the next tick is not your friend. In options, this can happen when implied volatility is shifting quickly or when the underlying is moving through strikes. You post a limit to sell an option, you get filled, and then the option keeps going up because vol is rising. You weren't "unlucky." You were providing liquidity to someone who wanted out.

This doesn't mean limit orders are bad. It means they are not magic. Smart execution is understanding that every order is a bet about microstructure:

> *Am I demanding liquidity, or supplying it? And what might the other side know?*

If you're consistently getting "great fills" on your limit orders, it's worth asking an uncomfortable question: are you being paid because you're providing liquidity, or because you're being picked off?

Choosing instruments: liquidity is a strategy choice, not a constraint

Volatility traders have a temptation to shop for "better pricing" in less liquid corners: weekly expirations with wider spreads, obscure strikes, thin VIX options series, small single-name options with juicy implied vol. Sometimes there's edge there. Sometimes there's just a liquidity premium you can't actually harvest after costs.

A practical definition:

> *A liquid instrument is one where you can be wrong and still get out without donating the account.*

When selecting what to trade, look for signs that you're not going to be trapped:

- tight spreads relative to option price,
- consistent displayed size and depth,
- high open interest (so you're not the only person in the room),
- healthy volume across strikes and expirations,
- stable behavior during the times you expect to trade (open, close, major events).

The subtle point: volume alone doesn't guarantee liquidity. Some instruments trade a lot but still have poor depth at the moments that matter. Liquidity is not how much trades. Liquidity is how much you can trade without getting punished.

Execution in volatility products: your worst fills often happen when your signals are ``strongest"

Many volatility strategies are built to react to spikes: sell vol when it's high, buy protection when it's cheap, roll when term structure changes, hedge when gamma rises. The irony is that those moments-when your signal is flashing brightest-are also when spreads widen and fills deteriorate.

So your strategy may be structurally chasing the worst execution moments. That's not a fatal flaw. It just means you need an execution plan that is part of the strategy, not an afterthought.

Two practical adaptations:

- *Stagger entry and exit*: scale in/out rather than trying to do everything at the worst tick.
- *Pre-define "good enough"*: decide in advance what price improvement is realistic so you don't waste time chasing pennies while the market runs away.

A surprising amount of execution skill is emotional management disguised as process: resisting the urge to "get it done" with a market order when you're stressed.

Rolling and rebalancing: when predictable flow becomes a tax

Roll periods concentrate activity. Lots of traders roll around the same times for the same reasons-expiry cycles, index rebalances, systematic strategies. Concentrated flow is where spreads widen and slippage increases, because the market can smell urgency.

If you always roll on the same day, at the same time, in the same size, you are teaching the market your habits. Markets love predictable habits. They monetize them.

A more robust approach:

- Roll earlier when liquidity is decent rather than waiting for the crowded door.
- Break rolls into smaller pieces.
- Use limit orders and be willing to let part of the order rest.

Rolling is not only a risk decision; it's an execution event. Treat it with the same respect you'd give a new trade, because the market charges you for treating it casually.

Measuring your own execution: keep score like a professional

You can't improve what you don't measure. The good news is that execution measurement doesn't require institutional tools to be useful.

Keep a simple log for your main instruments:

- quoted spread at the time you traded,
- where you filled relative to mid (improvement or slippage),
- whether the order was passive (resting) or aggressive (crossing),
- whether the market was calm or stressed.

Over time you'll see patterns. Maybe you're consistently paying too much at the open. Maybe you chase fills when volatility spikes. Maybe you get picked off when you place tight limits in fast markets. Those patterns are tradable. They are also fixable.

The goal isn't to become an execution scientist. It's to stop being an execution charity.

A realistic ``edge budget'': allocate profits to costs before you spend them

Execution costs are not random; they are semi-structural. That means you can budget for them.

If a strategy is expected to earn E per trade (or per month) before costs, you should mentally allocate a chunk of that to:

- spreads,
- slippage,
- hedging churn,
- and the occasional bad fill during volatility events.

Then ask a blunt question: after paying those bills, is there enough left to justify the risk?

This is where many paper strategies fail. They have a theoretical edge that is smaller than the real trading frictions. The solution is not necessarily "don't trade." It's often:

- trade fewer times with higher conviction,
- use more liquid strikes/expiries,
- size down so you don't move the market,

- simplify structures so you can execute cleanly.

In other words: make your edge *robust* to reality.

Actionable takeaway: adopt the ``limit-first, liquidity-first" rule

For every volatility trade, apply two defaults unless you can explain why you're breaking them:

1. *Limit-first*: start with a limit order placed at a realistic price (often near mid but not fantasy-mid), and only get more aggressive in controlled steps.
2. *Liquidity-first*: prefer strikes and expiries where spreads are tight, open interest is healthy, and you can exit under stress without relying on heroic fills.

Then add one weekly ritual: compute your portfolio at liquidation marks and compare it to your mid-mark view. If the difference is large, you're not running a better strategy-you're running a strategy that depends on the kindness of spreads. Kindness is not a stable market regime.

7.6 The Art of Rolling Positions

Rolling sounds like housekeeping: take out the near-dated contract, replace it with a later one, keep the exposure. In practice it's closer to surgery. You're cutting out one risk profile and stitching in another, usually while the patient is awake and the room is full of noise. Done well, rolling is how you keep a good idea alive without letting time, gamma, and market microstructure turn it into a different animal. Done badly, rolling is how you turn "managed trade" into a slow-motion margin call.

The first thing to admit—out loud, to yourself—is that a roll is not neutral. Even if the delta looks similar before and after, you've changed the trade's *geometry*. You've swapped time decay, convexity, and sensitivity to implied volatility. You've also swapped your exposure to specific calendar landmines: settlement quirks, dividend dates, central bank meetings, earnings, index rebalances, and those weird mornings where the opening print feels like a practical joke.

Rolling is where discipline meets storytelling. Your brain wants continuity: "I'm still in the trade." The market doesn't care about continuity. It only cares what you hold *now*.

Two kinds of rolls: maintenance and rescue

Most roll decisions are secretly one of two things.

A *maintenance roll* is boring in the best way. Your thesis is intact, the position is behaving, and you're rolling because time is passing and you want to maintain an exposure (or avoid a nasty part of the risk curve). This is how systematic strategies survive: they treat rolling as a scheduled part of the machine.

A *rescue roll* is emotionally spicy. The position is wounded, you don't like the P&L, and the roll is a way to postpone taking a loss. Rescue rolls can be legitimate, but they are where traders lie to themselves most often. The market can smell denial. It charges extra.

A simple litmus test:

> *If this position didn't exist today, would I put it on fresh at the rolled strikes and expiry?*

If the answer is no, the roll is probably a rescue mission disguised as "management."

Rolling changes your Greeks, even if you pretend it doesn't

Every expiration lives on its own risk planet. Near-dated options have higher gamma and more violent behavior around spot moves. Farther-dated options tend to be more vega-heavy and less twitchy day-to-day, but they can be more sensitive to changes in the implied volatility regime.

When you roll from front month to next month, you typically:

- reduce near-term gamma risk (less acceleration of pain),
- change theta (time decay per day shifts),
- change vega (sensitivity to implied vol levels often increases with maturity),
- change skew exposure (the surface can differ across maturities).

So "rolling" isn't a single action. It's a trade where you're buying and selling Greeks. If you don't name which Greek you're trying to change, you're letting the calendar change it for you.

A useful mental model: the expiry you choose is not just a date-it's a *risk thermostat*. Shorter-dated structures can generate quicker decay but demand more precise risk control. Longer-dated structures give you time but can accumulate premium and vega exposure in ways that behave differently under regime shifts.

The roll decision framework: thesis, mechanics, and market

A roll deserves the same three questions every new trade deserves. They're simple, but they're the ones people skip when the clock is ticking.

Has the thesis changed? If your original reason for the position is gone, rolling is just extending a mistake. Close it. There's no moral victory in keeping a trade alive after its idea died.

Have the mechanics changed? Time decay, gamma, settlement rules, assignment risk, and liquidity can all worsen as expiry approaches. Sometimes you roll not because you *want* to, but because you don't want to be forced into a worse set of mechanics.

Has the market changed? If implied vol has moved, skew has moved, or liquidity has deteriorated, your roll is happening in a different environment than the one you entered. Rolling into a volatility spike can be brilliant or suicidal depending on whether you're buying or selling premium. Rolling into an event week can change the trade even if the strikes look the same.

Notice what's missing: "How do I feel about being down?" Feelings are real, but they are not inputs.

The hidden cost of rolling: you pay the spread twice

Rolling means you trade two legs: you close one position and open another. That often means you cross the spread twice, and you do it in a moment when liquidity can be worse than average (because other people roll around the same times).

This is why rolling can quietly destroy a strategy that looks great on paper. If your edge is small, paying the spread twice on schedule is like buying a gym membership you never use-you're funding someone else's lifestyle.

A practical execution rule: whenever possible, roll as a *single spread order* (a calendar spread, diagonal, or roll combo) rather than two separate trades. This can:

- reduce legging risk (the market moves between your close and open),
- improve pricing (market makers price the relationship),
- make your true roll cost visible.

It doesn't guarantee a better fill, but it gives you a fighting chance.

Don't roll into landmines: the calendar matters

There are weeks when the market is basically normal, and there are weeks when the market is a live wire. Rolling is how you accidentally grab the live wire.

The classic landmines:

- major macro events (central bank meetings, CPI),
- big earnings weeks (if you trade single names or exposures tied to them),
- index rebalances and quarter-end flow,
- settlement quirks in certain products.

The point isn't to avoid events forever. The point is to roll *with awareness*. If your strategy is designed to harvest premium, rolling into an event week can be fine if you are being paid enough and your risk is defined. If your strategy relies on stability, rolling into an event week is like moving your furniture during an earthquake drill.

A trader's version of "measure twice, cut once":

> *Before rolling, check the next expiry's event calendar as if you were a landlord checking a tenant's credit.*

Assignment risk: the roll that can be forced on you

If you are short American-style equity options, you have a special risk that doesn't care about your schedule: early assignment. Assignment is the process where your short option is exercised and you wake up with stock (or short stock) you didn't intend to hold. It's not predictable in a clean way because assignment is driven by the long holder's incentives and the operational reality that you can't see.

Assignment becomes more likely when an option is deep in-the-money and close to expiration, and it has famous "hot zones" around dividends for calls. The key point for rolling is behavioral:

> *If you wait until the last minute to roll short American-style options, you may discover the position rolled itself—into stock.*

That isn't always catastrophic, but it can be operationally messy, margin-intensive, and emotionally destabilizing—

exactly the wrong mix when you're trying to manage risk calmly.

So for short American-style options, rolling is partly about reducing the probability of being forced into an unintended position at an inconvenient time. Sometimes the most professional roll is the one that feels "early," because it avoids the week where assignment risk and liquidity risk start holding hands.

VIX options: the settlement you *think* you're trading is not always the one you get

VIX options are a different creature. They are generally European-style (no early exercise), which removes a lot of assignment drama. But they introduce their own version of surprise: settlement is based on a special opening quotation (often referred to as VRO), derived from opening prices/quotes of the SPX options used in the VIX calculation. That settlement print can differ meaningfully from the VIX level you saw the day before, or even from what you see intraday.

For rolling, this matters in two ways:

- *Expiry is not just expiry*: the settlement morning has its own behavior. If you're holding into settlement, you're exposed to the difference between "VIX I watched" and "VIX that settles."
- *Roll timing changes risk*: rolling before settlement can reduce exposure to settlement basis risk, but it also means you're trading when liquidity may be uneven.

The practical advice is not "avoid VIX options." It's "respect the contract." If you use VIX options for tail hedging or directional vol views, rolling is part of ensuring your hedge behaves like a hedge and not like a trivia question.

VIX futures: rolling the curve, not just the contract

VIX futures add a different twist: term structure. When you roll VIX futures, you're not only changing time-you're trading a curve that can be in contango (farther months higher) or backwardation (near months higher). That curve shape is not background noise. It can dominate returns.

If the curve is in contango, rolling a long futures position can create a negative carry headwind because you're selling a cheaper contract and buying a more expensive one. If the curve is in backwardation, the roll can help you. The roll is literally a P&L event.

So a VIX futures roll is never "just maintain exposure." It's "maintain exposure while accepting (or harvesting) the curve's carry." If you don't want curve carry, you need a different instrument or a different structure.

One more practical constraint: large positions can be subject to position accountability regimes and liquidity realities near expiration. Even smaller traders should adopt the professional habit: don't wait until the last minute when the roll becomes crowded and the market can price your urgency.

The most common rolling mistake: confusing time with forgiveness

Time feels like forgiveness because it gives you more opportunities for the market to come back. But time also changes the trade you're holding, and it can increase the total premium and risk you've put at stake. Rolling a losing short-premium position forward can be a rational repair. It can also be a way to turn a manageable loss into a larger, slower loss with the same underlying vulnerability.

A useful phrase to keep yourself honest:

> *A roll can change your odds, but it cannot rewrite your edge.*

If your edge is gone-if the market regime changed, if skew repriced, if liquidity deteriorated-rolling is not a reset button. It's a new trade that deserves a new sizing decision, not an emotional continuation of the old one.

How to roll without turning it into a coin flip

A clean roll process has three parts: preparation, execution, and verification.

Preparation means deciding what you want the roll to accomplish in Greek terms. Are you trying to reduce gamma risk? Extend theta harvesting? Reduce settlement risk? Maintain a hedge?

Execution means choosing how to trade the roll so you minimize spreads and leg risk. Often that means trading the roll as a single order, using limits, being patient, and avoiding the most chaotic minutes of the day unless you have to.

Verification means checking what you actually ended up with. After the roll, confirm:

- your new delta/gamma/vega profile is what you intended,
- your exposure to events and settlement is understood,
- your liquidity and margin usage are acceptable under stress assumptions.

A shocking number of "mysterious" losses come from rolls that accidentally changed the strategy: wrong strikes, wrong maturity, wrong product, wrong event week. Rolling is a time when operational mistakes become financial mistakes instantly.

Actionable takeaway: write a roll ticket with intent, not just strikes

Before you roll any position, write a two-line roll ticket for yourself:

1. *Intent*: "I am rolling to ______________" (reduce gamma, avoid settlement, maintain hedge, harvest premium, etc.).
2. *Red line*: "If the roll costs more than __________ (in spread/price), or if liquidity is thin, I will __________" (scale, delay, reduce size, close instead).

Then execute the roll as if it were a brand-new trade, because it is. The market doesn't grade you on continuity. It grades you on what you hold after the roll—and whether you can still sleep while holding it.

Chapter 8

The Professional Edge: Data, Discipline, and Psychology

You have the strategies and the theory, but strategies do not trade themselves-people do. The difference between a lucky amateur and a lasting professional isn't usually IQ; it is infrastructure. This chapter explores the invisible scaffolding that supports profitable trading: the quality of your data, the honesty of your backtests, the strictness of your rules, and the resilience of your mindset. We will turn your trading from a stressful hobby into a disciplined business.

8.1 Data Basics: Getting the Right Numbers

A stock trader can get away with a little ignorance. A volatility trader can't.

If you buy 100 shares of an S&P 500 ETF, the world is mostly one-dimensional: price goes up, price goes down, you live or die by the tape. But if you trade volatility—whether you think of it as VIX, SPX options, variance, skew, or "premium harvesting"—you're trading a *map* made from thousands of tiny prices that each have their own rules. You're not just asking, "Where is the underlying trading?" You're asking, "How expensive is protection today, at each strike, for each expiry, under each micro-condition of liquidity and order flow?"

The uncomfortable truth is that most blown-up volatility accounts weren't killed by a bad idea. They were killed by bad numbers: stale quotes, mis-tagged expirations, missing corporate actions, a futures roll modeled with fantasy fills, or an implied-vol series built from yesterday's option chain because that's what the free API happened to serve. When your strategy's edge is measured in a few basis points per day (and your tail risk is measured in *years of gains* disappearing in an afternoon), "close enough" data is the same thing as wrong data.

Picture a friend showing you a clean equity curve from "selling VIX"—steady gains, tiny drawdowns, a Sharpe ratio that looks like it was printed by a marketing department. You ask one question: *Which VIX are you using?* Spot VIX? Front-month VIX futures? A VIX ETP? Synthetic VIX from SPX options? They blink. That blink is the problem.

Volatility has an ecosystem. Every piece of it references a different instrument with different mechanics, trading hours, and settlement quirks. In equities, confusing AAPL with MSFT is a rookie mistake. In volatility, confusing *spot VIX* with *VIX futures* is a professional-grade error that still shows up in "research" every day. Spot VIX is a *calculation* derived from SPX option prices; VIX futures are *tradable contracts* with their own supply/demand dynamics. They talk to each other, but they are not the same language.

So, what does "getting the right numbers" actually mean? It means building your data stack around what volatility trading truly consumes: not a single price series, but a triangulation.

The Holy Trinity of vol data looks like this:

- *The underlying spot price* (OHLC and preferably intraday): the ground you stand on.
- *Options chain snapshots*: the full menu of implied volatilities (the surface), not just one headline number.
- *Futures term structure*: the curve that quietly taxes you (or pays you) through roll yield.

Miss one of those, and you're trading a shadow. You might still make money for a while. Shadows can be profitable. They're also famously hard to measure when the sun moves.

The underlying spot series is the part everyone thinks they already have, and that's precisely why it's dangerous. "I have SPY daily candles" is fine if you're swing trading. But volatility strategies are often *path-dependent*: how the market *traveled* matters, not just where it ended. Two days can close at the same price and produce radically different outcomes for options.

Consider a simple, brutally common vol trade: short 30-day implied volatility through an index option position. Your P&L is sensitive to realized volatility (the market's actual movement), which isn't captured by the close alone. A day that whipsaws 2% down and 2% up but finishes flat is not "a flat day" to your short vol book. It's a day where gamma and hedging costs can chew you up like a paper shredder.

Actionable implication: for volatility work, daily OHLC is the bare minimum, but intraday bars (even coarse ones like 5-minute or 15-minute) can be the difference between an honest backtest and a bedtime story. You don't necessarily need tick data; you *do* need enough resolution to detect whether your edge survives the kind of intraday noise that options pricing actually cares about.

The second piece—options chain snapshots—is where most data setups quietly fail.

A stock chart is a movie. An options chain is a *crowd photo*, and you need many crowd photos over time. You don't want just "the IV"; you want the *surface*: implied volatility across strikes (moneyness) and across expirations (tenor). That surface is where skew lives. It's where crash insurance gets priced. It's where the market tells you, in dollars, what it fears.

Here's the problem: most retail-grade data services don't actually give you chain *snapshots*. They give you a chain when you ask for it. That sounds the same until you try to do research. A chain "when you asked" is contaminated by your query time, your latency, and the vendor's caching logic. It's like trying to measure ocean tides by dipping a cup of water at random hours and calling it a dataset.

A proper snapshot is time-stamped and complete: all strikes, greeks (or enough to compute them), bid/ask, last, volume, open interest, and a clean identifier for the exact contract (including whether it's AM- or PM-settled, weekly vs monthly, and the precise expiration date). You want it stored so you can say, *At 10:30 a.m. ET on this date, this is what the market offered*. That one sentence is the difference between research and reconstruction.

Bid/ask matters more in volatility than it does in stock trading because the thing you're trading is often *the spread itself*: small expected returns harvested repeatedly. If your backtest prices every option at the mid, you've built a strategy that trades at a price that doesn't exist. That can be fine as a theoretical ceiling. It's lethal as a plan.

A good habit: treat the bid/ask like gravity. You can jump, but you can't ignore it. If you wouldn't get filled at the mid in size during normal markets, you should not assume the mid in backtests. And if your strategy relies on getting the mid during stress, you should assume it breaks exactly when you need it most.

Chain snapshots also force you to learn an unglamorous truth: options are not one market; they are many small markets glued together by arbitrage *when liquidity allows*.

Some strikes are liquid. Some are a desert with a mirage of a last price from two hours ago. Weeklies can be lively; far-dated can be sleepy; deep OTM puts can be liquid in a crash and dead in a calm. If your vendor serves you "last" as if it were truth, you'll end up building implied volatility surfaces out of ghosts.

Practical rule: for research, prefer *quotes* (bid/ask) over trades (last). Trades are evidence that something happened; quotes are evidence of what you can do *now*. Volatility trading is execution-sensitive—especially when the strategy is short convexity and needs to get out fast.

And remember: options change *identity*. Contracts expire; strikes get added; weeklies appear; special expirations get listed around events. If your dataset can't keep track of contract metadata cleanly, your model will accidentally compare apples to an apple sticker on a banana.

The third pillar—the futures term structure—is where volatility strategies either become elegant or become expensive.

If you've ever held a VIX futures-linked product and wondered why it bleeds even when "VIX stays the same," you've met the term structure. VIX futures trade in a curve: different months priced differently. When the curve slopes upward (contango), rolling from front-month to next-month tends to cost you. When it slopes downward (backwardation), the roll can pay you.

This is not academic. It's the rent you pay to stay in the trade.

And here's the trap: many traders model futures exposure as if it were a single continuous series with no frictions. They take a front-month contract, stitch it to the next, and call it "VIX futures." That might be acceptable for a rough macro picture, but it is not sufficient for strategies where the *roll schedule* is the engine of returns (or losses). If you trade anything that maintains a constant maturity exposure, you must model *how* that constant maturity is maintained—what you sell, what you buy, and when.

You need daily settlement prices for each listed contract, plus the contract specs (multiplier, tick size, settlement type), and you need a consistent roll rule that matches what you'd do live. A two-day difference in roll timing can materially change results during volatile periods, because the front of the curve can move like a shaken soda can.

Even if you never trade a single futures contract, you still need term structure data because it affects the pricing of volatility across time. The market's expectation of future volatility is embedded in that curve, and many strategies are, whether they admit it or not, making a bet on the slope.

Now for the part that makes professionals quietly nod: volatility data is *multidimensional*, and dimensionality creates opportunities for self-deception.

When someone says, "I trade volatility," ask: which maturity? which strike region? which underlying? which trading hours? which settlement? are you using SPX or SPY options? Are you mixing European-style and American-style options in the same model? Each answer is a different dataset with different edges and different landmines.

Even the calendar isn't as simple as it looks. Options don't decay linearly through time, and weekend/holiday effects matter. A "30-day" option on a Friday afternoon has a different feel than a "30-day" option on a Monday morning, because the market is pricing the next weekend, the next macro events, and the next gap risk. If your data doesn't have clean timestamps and you blur everything into end-of-day closes, you're smearing important information into a single pixel.

There's also the quiet issue of *trading hours mismatch*. SPX options have their own liquidity rhythms, and futures trade nearly around the clock. If you compute signals using a futures price at 2:00 a.m. and then assume you entered an options trade at the 4:00 p.m. close at the same implied vol level, you've invented a teleportation device. Research loves teleportation. Markets don't.

A clean way to avoid this: define a daily "decision time" and enforce it across all datasets. For example: signals computed from data available at 3:55 p.m. ET, trades assumed at 4:00 p.m. ET, and only using quotes observed before entry. That kind of boring discipline is where the edge hides, because it prevents your model from accidentally peeking into the future.

Data quality isn't just about accuracy. It's about *definitions*. Volatility is famous for numbers that sound the same but mean different things.

Realized volatility is what happened. Implied volatility is what the market charges for protection. VIX is a particular implied volatility measure derived from SPX options with specific rules. A broker "IV" field might be at-the-money implied vol, or it might be a weighted blend, or it might be a proprietary estimate. Two services can both label a column "IV" and be describing different planets.

If you're going to compute your own implied vol surface (and serious vol research eventually pushes you there), you'll need consistent inputs: option prices, a risk-free rate proxy (often simplified), dividend assumptions for equity underlyings, and a clear pricing model choice. For index options, details like forward calculation and rate assumptions matter less than people fear on calm days and more than they expect in stress, when small model mismatches become large P&L mismatches.

A useful mindset: every data field is a claim. Ask what it is claiming, how it was computed, and under what conditions it breaks.

Costs deserve their own paragraph because costs are where volatility strategies go to die quietly.

A typical stock backtest might survive with a flat commission assumption. Volatility strategies often require explicit modeling of:

- bid/ask spreads that widen in stress,
- slippage that increases with urgency,
- option exercise/assignment mechanics (if applicable),
- exchange and regulatory fees (small, but real),
- margin and financing assumptions (especially for short options),
- futures roll mechanics (the curve is a cost center).

When someone says, "It works before costs but not after," what they often mean is, "My data assumed a world where liquidity is free and fear is orderly." Fear is never orderly. Fear is when spreads yawn open, fills get partial, and your strategy's expected edge tries to hide under the couch.

So the right numbers include the right *implementation* numbers. If you're trading a strategy that earns pennies per day but is exposed to a steamroller, you can't afford to pretend execution is perfect.

Where do you actually get this stuff without lighting your budget on fire?

Start by separating two use cases: *learning and monitoring* versus *research-grade backtesting*. For learning and monitoring, broker platforms and decent retail data providers can be enough. The goal there is situational awareness: current term structure, current skew shape, approximate implied vol levels, and enough historical context to recognize regimes.

For research-grade work, you need point-in-time history. That means historical options chains captured as they existed then, not reconstructed later with survivorship and convenience baked in. It's more expensive because it's more honest.

There's a middle path many traders miss: you can build your own snapshot archive going forward. You don't need ten years of history on day one. If you start capturing daily chain snapshots now—even once per day at a fixed time—you're building an asset that compounds. After a year, you have a year of clean, consistent snapshots for your exact decision time. After three years, you've got something you can't download for free, and you understand it because you built it.

If you're serious, automate it:

- Pick a decision time (e.g., 15:55 ET).
- Pull the underlying spot data (price and ideally intraday bars).
- Pull the full options chain (bid/ask, IV, OI, volume, greeks if available).
- Pull the futures curve (all listed expiries you care about).

- Store raw data *and* cleaned data (so you can audit your cleaning).

The raw archive is your insurance policy against your future self discovering that one of your cleaning steps was "creative."

Cleaning is where the "free money" myths usually collapse.

Vol data arrives messy. Duplicated strikes, bad quotes, crossed markets, missing bids, contracts that exist but don't trade, timestamps that don't align, and fields that occasionally go haywire. You don't solve this with hope; you solve it with rules.

A few rules that pay for themselves:

- Throw out options with zero bid when modeling sell-side fills unless you have a realistic way to execute (you often don't).
- Use bid for sells and ask for buys in pessimistic backtests; use mid only as a best-case sensitivity check.
- Filter out absurd implied vols produced by tiny option prices (a $0.01 option can imply "infinite" vol depending on assumptions).
- Keep contract identifiers strict; never merge contracts just because strikes and expirations "look similar."
- Log every cleaning action so you can reproduce results.

None of this is glamorous. That's why it works. Most traders would rather tweak an entry signal than build a data pipeline that tells them the uncomfortable truth. The market rewards the person willing to be bored in exactly the right way.

Finally, don't ignore the *context* data that turns a dataset into a decision system.

Volatility is sensitive to events: CPI releases, Fed meetings, earnings (for single names), quarterly expirations, month-end flows, and sudden risk-off headlines. You don't need a PhD-level event model, but you do want an events calendar in your dataset so you can label regimes. Otherwise, you'll look at a spike in implied vol and treat it as random noise when it was the market repricing a known binary event.

Regime labeling is also how you stop overgeneralizing. Many vol edges exist only in certain environments: calm contango months, stressed backwardation, post-crash normalization, pre-event premium inflation. If your data isn't rich enough to distinguish those environments, your strategy will look like it works "on average" and fail precisely because the average is not where you live.

The punchline is simple: volatility trading is not harder because the math is scary. It's harder because the instrument is a moving target and the data is a three-headed beast. If your numbers are fuzzy, your conclusions will be confident and wrong—the most expensive combination in finance.

Actionable takeaway: pick one volatility product you care about (e.g., SPX options or VIX futures) and build a one-week "data audit" before you build a strategy: collect a daily underlying price series, one fixed-time options chain snapshot per day, and a daily futures curve (if relevant). Then answer three questions in writing: (1) exactly what instrument am I trading, (2) exactly what timestamp are my signals allowed to use, and

(3) exactly what price (bid/ask/mid) am I assuming I can execute. If you can't answer those crisply, the strategy isn't under-researched—it's undefined.

8.2 Backtesting: Avoiding False Confidence

A backtest can make you feel like a genius in under five minutes.

You load some historical data, write a few lines of logic, and the equity curve starts climbing like it has inside information. Your brain does what it's built to do: it connects dots, smells patterns, and starts mentally spending profits you haven't earned. The market, meanwhile, is patiently waiting for you to confuse a simulation with a skill.

Volatility strategies are especially good at producing this kind of false confidence, because they often monetize something that looks stable right up until it isn't. Short-vol strategies can print small gains so consistently that almost any backtest window feels "validated." Long-vol strategies can look like slow-motion failure until you accidentally choose a sample that includes a crisis and suddenly you've "discovered convexity." If you don't interrogate the backtest like an auditor, it will happily tell you whatever story you want to hear.

The real purpose of testing isn't to prove you're right. It's to find out *how* you're wrong before the market charges you tuition.

Start with the most common hidden sin: using information you didn't have yet. It rarely looks like cheating. It looks like convenience.

Lookahead bias is any situation where your strategy's decision at time t quietly depends on data from time $t+\epsilon$. Sometimes it's blatant ("buy at the close using the close" without accounting for the fact you can't know the close until it happens). Sometimes it's subtle (using an "end-of-day implied vol" field that is actually computed using the settlement process that occurs after the last chance you had to trade). And sometimes it's downright sneaky: your dataset vendor might deliver a "daily" options snapshot that is actually a blend of prints and quotes collected throughout the day. Your code treats it as a single moment in time. Your backtest gets to time-travel.

Volatility trading makes this worse because the inputs are multidimensional. One timestamp mismatch is enough to manufacture edge. If your signal is built from SPX option implied vols stamped at 4:15 p.m. but your execution assumes fills at 3:55 p.m., you've created a free forecast of the last 20 minutes of the day. In calm markets, that might not matter. In stress, those 20 minutes can contain the entire move.

Here's a quick smell test: if your backtest assumes you enter on the same bar that generated the signal, it's guilty until proven innocent.

Look at the two curves and ask a question that feels almost insulting: *Which one looks like a market?* Markets don't move in polite stair-steps. They stall, they lurch, they punish you for being early and mock you for being late. If your equity curve looks like a savings account with better branding, the first thing to suspect is not "skill." It's "timing that wasn't actually available."

Fixing lookahead bias is less about cleverness and more about courtroom-grade procedures:

- Define a *decision timestamp* and enforce it across all inputs.
- Only use fields that exist *at or before* that timestamp.
- Separate signal generation from execution (even one bar later is more honest than same-bar fills).
- Use bid/ask logic consistent with what you're doing (selling at bid, buying at ask) unless you can justify and measure improvement.

If this feels strict, good. Strict is what keeps your backtest from quietly upgrading you into a time traveler.

The next trap is survivorship bias, and it's cruel because it flatters you with "clean" data.

Survivorship bias happens when the dataset only includes the winners that survived long enough to be convenient. In stocks, that's companies that didn't delist. In volatility and derivatives work, it shows up differently: contracts roll and expire, products get shut down, index constituents change, and the liquidity you see today wasn't always there. If you test a strategy on a universe that assumes today's availability and today's tight spreads existed forever, you're studying a market that never existed.

A volatility example that bites hard: backtesting a strategy using modern, liquid weekly options and assuming those same instruments and spreads existed in earlier years. You'll often "discover" an edge that is actually the shadow of liquidity improvements over time. The market didn't become easier; the market became easier to simulate.

A good dataset doesn't just contain prices. It contains the *truth of what could be traded* at the time: which contracts existed, what their spreads were, and whether size could be executed without crossing the earth's crust.

Then there's the crowd favorite: overfitting. The strategy that wins the past because you trained it to.

Overfitting isn't just "too many parameters." It's the broader habit of trying a lot of things and only remembering the one that worked. If you test 200 variations of a volatility strategy—different entry thresholds, different maturities, different stop rules, different roll schedules—it's not surprising when one looks amazing. It would be surprising if none did. That's not edge; that's statistics doing what statistics does: producing an apparent signal when you sift hard enough.

Volatility strategies are particularly overfit-prone because the market has regimes. A rule that performs brilliantly in contango-heavy years can look heroic, until the curve flips and the rule becomes a professional way to donate money. You can accidentally tune a strategy to the *sequence* of regimes in your sample, not to a durable relationship.

The dangerous part is psychological: an overfit backtest is not a random mess. It often looks *beautiful*. It lines up with a story. "It sells insurance when fear is overpriced." "It buys protection when the market is complacent." You can tell that story at a dinner party and everyone nods. That's how overfitting recruits allies.

A practical defense is to treat every knob you add as a tax. Each new parameter is a new way to be fooled. If you can't explain why a parameter exists *without referencing the backtest result*, it's probably a souvenir from curve-fitting.

Validation is where most retail backtests commit their final crime: the single heroic split.

People love the idea of "in-sample" and "out-of-sample" as if one clean divide separates science from superstition. The problem is that research rarely stops after the split. You test, you adjust, you re-test, you adjust again. Every time you peek at the results and tweak the rules, you've dragged information from the future into the past, even if you never touched the out-of-sample window directly. The process itself becomes a selection machine: it selects what looks best *given the history you happened to see.*

A more realistic approach is walk-forward thinking: multiple slices of time where you pretend you live there, make choices based on the past, and then observe how those choices behave in the next slice. It feels slower and less satisfying, because you don't get one triumphant equity curve. You get a stack of smaller, imperfect curves—which is exactly the point. The market doesn't give you one long, uninterrupted "track record"; it gives you a series of seasons where the rules of pain change.

If you want one mental model, use this: a strategy that only works when you know which years are coming is not a strategy. It's a biography.

Volatility adds a special kind of landmine: the backtest that assumes rebalancing is frictionless.

A lot of vol strategies rely on frequent adjustment: rolling options, delta hedging, maintaining target maturity, keeping exposure within limits. On paper, this looks like hygiene. In practice, it's where the P&L leaks out.

Even a modest rebalance schedule interacts with liquidity and spreads in non-linear ways. When markets are calm, rebalancing is cheap. When markets are violent, rebalancing is expensive precisely when it matters most. If your backtest models costs as a fixed number per trade, it will understate the cost of trading *during stress*—which is like underestimating the cost of water *in a desert*.

A robust backtest will include cost models that widen with volatility, or at least stress scenarios where spreads are multiplied (2x, 3x, 5x) during high-VIX regimes. If your edge disappears under a cost stress that is plausible, you've learned something priceless without paying the market to teach it.

Another quiet source of false confidence: using the wrong price for the job.

Marking to mid is the classic example. The mid is not a tradable price; it's a negotiation that hasn't happened. Mid-marking can be a useful *benchmark* if you treat it as a best-case. But when a strategy's profit per trade is small—which describes a lot of volatility harvesting—mid-marking can turn a negative expectancy strategy into a positive expectancy illusion.

There's also the fill-timing fantasy: assuming you trade exactly at the settlement price. Settlement is often a calculation, not a fill. It can be a helpful reference point, but it's not your broker saying, "Sure, take as much as you want at that number."

A simple discipline that helps: run two versions of the backtest.

- *Optimistic*: mid fills, minimal slippage, low fees.
- *Defensive*: bid/ask fills, realistic slippage, higher fees, spread widening in stress.

If the strategy is only alive in the optimistic world, it's not a strategy; it's a wish.

Now the uncomfortable part: even a perfectly honest backtest can still be misleading, because volatility returns are not polite.

Many vol strategies have return distributions that are skewed and fat-tailed: lots of small outcomes and a few enormous ones. That creates two problems. First, a short sample can flatter you. Second, a good-looking average can hide an unacceptable worst case.

This is where "false confidence" becomes more than a research issue. It becomes a risk issue. If your strategy sells convexity, the backtest may show steady growth for years and then one air pocket that wipes out the whole story. If your strategy buys convexity, the backtest may show long stretches of bleeding that look unacceptable until the one period where it pays off massively.

So a realistic validation has to answer questions that feel less fun than CAGR:

- What is the worst drawdown, and how fast does it happen?
- How sensitive is performance to a handful of days?
- Does the strategy depend on one crisis month to look smart?
- What happens if you shift entries by one day? By one hour?

A strategy that dies when you nudge the timing is telling you a secret: it was never robust; it was just well-aligned with the past.

One of the most effective realism checks is "damage testing": deliberately injure your backtest and see if it survives.

Here are injuries that mimic real trading life:

- Randomly delay entries by 0–1 trading days.
- Randomly worsen fills by a few cents or a small percentage of option premium.
- Remove a small fraction of the best days (because you might miss them).
- Increase costs during high-volatility regimes.
- Cap liquidity: limit position size or force partial fills.

If performance collapses under mild injuries, the edge is fragile. Fragile edges are the ones that look best in a backtest and feel worst in a live account.

You can also perform a "parameter neighborhood" test: if the strategy uses a threshold (say, enter when some implied-vs-realized measure exceeds X), examine results for $X - 10\%$, X, and $X + 10\%$. Real edges tend to be lumpy but present across a neighborhood. Overfit edges tend to be a needle: move it slightly and it vanishes. The market rarely rewards exactness. It rewards robustness.

There's a social version of overfitting that's worth calling out because it's everywhere in volatility content: borrowing a strategy template and then optimizing it to your data.

Someone posts "sell options when IV rank is high." You test it. It's okay. Then you tweak: maybe not IV rank, maybe IV percentile; maybe 30 DTE, maybe 45; maybe roll at 21 days, maybe 14; maybe add a filter on term structure slope; maybe exclude earnings; maybe include only Tuesdays. Eventually you produce a monster backtest and feel like you improved the idea.

What you often improved is your ability to tailor the past to your preferences.

The antidote is pre-commitment. Decide in advance how many variants you will test and what constitutes success. Write it down. It sounds overly formal for a personal trading project, but it protects you from the most dangerous person in your research process: you, feeling clever at midnight.

A final pitfall that trips up vol traders: confusing instrument exposure with concept exposure.

You might think you're testing "short volatility," but you're actually testing "short VIX futures roll in contango," or "short SPX put skew," or "short gamma with daily hedging," each of which has different behavior in crises. If the backtest doesn't decompose where P&L comes from—carry, roll yield, changes in implied vol, realized vs implied spread, skew changes—you can end up believing you've tested one thing when you've tested another.

When a strategy fails live, it rarely fails politely. It fails by revealing what it truly was all along.

Actionable takeaway: take your favorite volatility strategy backtest and run a "Reality Checklist" before you trust it: (1) enforce a single decision timestamp and shift execution at least one bar later, (2) re-price trades using bid for sells and ask for buys, (3) widen spreads and slippage during high-volatility periods, (4) rerun with entries randomly delayed by 0–1 days, and (5) verify performance doesn't hinge on a tiny number of extreme days. If the curve only looks good when the world is frictionless and perfectly timed, the backtest isn't revealing an edge—it's revealing a loophole.

8.3 Measuring Success Beyond Simple Returns

A volatility strategy can be "profitable" the same way a casino game can feel profitable: lots of small wins, a clean-looking bankroll chart, and one day where you discover what the word *limit* really means.

That's why simple return metrics—P&L, CAGR, even the classic Sharpe ratio—can be dangerously incomplete in volatility trading. They answer the question, "How much did you make on average?" while quietly dodging the question that matters: "How does this strategy *fail*?" Volatility strategies often fail in ways that don't show up in averages. They fail in the tails: rare, violent moves where liquidity thins, implied vol gaps, correlations snap together, and a position that looked "diversified" becomes one big bet.

It's not pessimism. It's job description.

Start with a mental picture that's almost too familiar: you sell premium for months, collect a steady drip of theta, and your account equity climbs with the confidence of a well-watered houseplant. Then a shock hits. The plant doesn't wilt. It combusts.

That pattern has a name: *negative skew*. In plain English, it means your strategy tends to deliver many small positive outcomes and a few large negative ones. You can have a high win rate and still have a terrible strategy. In fact, the higher the win rate in short-vol trades, the more suspicious you should become, because selling insurance is *supposed* to win most of the time.

Now add a second feature common to volatility returns: *kurtosis*, which is a fancy way of saying "fat tails." A normal distribution (the bell curve you learned in school) assumes extreme events are rare and nicely bounded. Volatility strategies live in a world where extremes show up more often than the bell curve promises—and they show up clustered, like bad weather systems.

So you don't just need metrics that reward performance. You need metrics that punish fragility.

The easiest way to feel the difference is to stop thinking in averages and start thinking in *survivability*.

If a strategy makes 2% a month but occasionally loses 40%, it may look great on a spreadsheet and feel horrifying in a real account. The math of recovery is brutal: a 40% drawdown requires a 66.7% gain just to get back to even. Your broker statement doesn't care about your Sharpe ratio when you're spending the next year digging out of a hole.

This is where volatility trading has an unglamorous truth: the most important number is often not your return. It's your *maximum drawdown*—the worst peak-to-trough decline you experienced. Max drawdown is not a perfect metric, but it has a virtue that most metrics lack: it speaks the language of pain.

And pain is operational. Pain changes behavior. Pain is where traders abandon their systems, widen their risk, revenge trade, or freeze at exactly the wrong moment. A strategy that is psychologically untradeable is financially irrelevant.

When you say the word "risk," most people picture volatility (standard deviation). Ironically, that's the one risk measure that can be least informative for volatility strategies.

Standard deviation treats upside and downside fluctuations as equally "risky." For a long-vol strategy, upside spikes are the whole point; for a short-vol strategy, small daily noise isn't the issue—the crash is. If your returns are skewed and fat-tailed, standard deviation can lull you into thinking you have control because most days look calm.

That's why professional evaluation leans on *drawdown-aware* metrics. They don't pretend risk is symmetric. They don't pretend the future is a bell curve. They focus on what can actually kill you.

A useful upgrade from Sharpe (for this world) is the *Calmar Ratio*. It's simple enough to remember and honest enough to respect.

Calmar is roughly:

$$\text{Calmar Ratio} = \frac{\text{Annualized Return}}{|\text{Maximum Drawdown}|}$$

It answers: "How much return did you get for the worst pain you had to endure?"

That framing matters because a volatility strategy can be engineered to show a pretty annual return by quietly carrying a massive drawdown risk. Calmar forces the trade-off into the spotlight. If two strategies both return 15% annualized, but one has a 10% max drawdown and the other has a 45% max drawdown, those are not cousins. They are different species.

Vol traders often discover an uncomfortable implication: the strategy with the best return can have the worst Calmar. And the strategy with the best Calmar is often the one that feels boring—because it's designed to survive the days that destroy accounts.

Before going deeper into metrics, it helps to see what "negative skew" looks like as a shape, not a slogan. The human brain understands pictures faster than formulas.

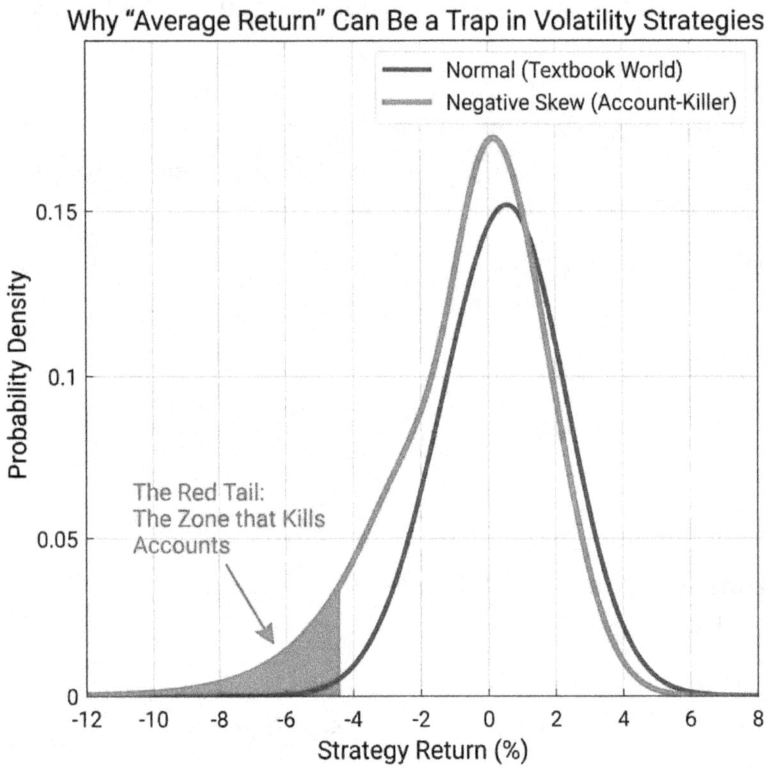

That red tail is the part that doesn't show up in a good mood. It shows up on the day you need your "risk management" to be more than a paragraph in your notes. A strategy can look like a steady income machine and still have a left tail so heavy it eventually collects all the profit back *with interest*.

Skew and kurtosis aren't trivia. They are a warning label.

Once you accept that tail risk is the story, evaluation becomes less about one hero metric and more about a small set of lenses.

One lens is *drawdown speed*. Two strategies can share the same max drawdown, but one takes 18 months to reach it in a slow bleed, and the other hits it in three days. Those are not the same experience. Fast drawdowns create forced selling: margin calls, de-risking mandates, the human impulse to "stop the pain." If your strategy can lose a year of expected gains in a week, you need position sizing and circuit breakers designed for *time compression*.

Another lens is *time under water*: how long you spend below your previous equity peak. This matters more than people admit. Spending months underwater is how traders abandon good systems. Spending years underwater is how traders stop trading. Volatility strategies, especially long-vol or hedged portfolios, can have long underwater periods even when they are ultimately valuable. If you don't measure that, you don't manage it.

Then there's the metric that traders love to quote and rarely respect: the Sharpe ratio.

Sharpe is "return per unit of volatility" (standard deviation). It's not useless; it's just often too easy to game in volatility strategies. If you smooth returns by marking to mid, underestimating costs, or using daily data that hides intraday shocks, Sharpe inflates like a balloon. Also, when you try many strategy variations and pick the best one, you're essentially shopping for the highest Sharpe. That's selection bias dressed as research.

A more honest way to treat Sharpe is as a *hypothesis*, not a trophy. High Sharpe invites a skeptical question: "Is this real, or did the process manufacture it?" If you've tried 50 variations and one has a Sharpe of 2.0, the correct emotional response is not excitement. It's suspicion.

Volatility strategies are the natural habitat of "Sharpe mirages" because they can create lots of small, autocorrelated gains. Autocorrelation means returns are not independent day to day, which violates a quiet assumption many people make when interpreting Sharpe. The metric looks better than the experience feels.

If your strategy's returns are lumpy and tail-driven, think of Sharpe like a speedometer in a snowstorm: it gives you information, but it can also encourage you to drive faster than conditions allow.

So what should you measure, concretely, when evaluating a vol strategy?

You want a dashboard that answers three questions:

- *How much does it make when it works?*
- *How does it lose when it loses?*
- *Can you actually stick with it long enough for the edge to show up?*

That suggests metrics like:

- Max drawdown and drawdown duration (pain and patience).
- Calmar ratio (return relative to worst pain).
- Worst week / worst month (time compression risk).

- Tail loss percentiles (e.g., 1st or 5th percentile daily return), as a simple proxy for "bad days."
- Exposure-adjusted return (did you earn return by taking hidden leverage?).

None of these is perfect. The point is redundancy. A single metric is a single point of failure, and volatility traders should be allergic to single points of failure.

Costs deserve to be inside the success metrics, not stapled on afterward.

A strategy with high turnover can look brilliant gross and mediocre net. In volatility, implementation drag has many flavors: spreads, slippage, roll costs, hedging costs, funding costs, and the occasional "you can't get filled where you thought" moment. Measuring success without costs is like measuring fitness without gravity.

One practical approach is to compute returns in layers:

- *Gross* (perfect fills, no costs) to understand the theoretical edge.
- *After realistic transaction costs* to understand tradability.
- *After stress costs* (wider spreads in high-vol regimes) to understand survivability.

If the edge only exists in the gross layer, you don't have an edge. You have a diagram.

There's also a subtle success measure that doesn't look like a metric until you've lived through a few drawdowns: *dependency*.

Dependency asks: does your strategy's performance depend on one type of event, one era, one product regime, one lucky crisis month? Many long-vol strategies look like dead money until the one period where they pay. Many short-vol strategies look like genius until the one period where they don't. If your "success" relies on a single chapter of market history, your backtest is a fragile narrative.

A simple way to expose dependency is to do contribution analysis: remove the best 5 days and worst 5 days and see what happens to the overall result. If removing a handful of days erases years of performance, you're not looking at a steady edge. You're looking at an option-like payoff. That's not automatically bad—options exist for a reason—but it changes how you size, how you fund, and how you judge success.

Volatility trading is full of strategies that are *correct* but unfundable, and strategies that are *fundable* but incorrect. Your success metrics should tell you which one you built.

One more angle that professionals obsess over (and retail traders often discover late): the difference between "strategy return" and "account return."

Your strategy might have a beautiful expected return, but your account experiences margin requirements, volatility of margin, and behavioral position changes. A short options strategy that looks modestly levered on calm days can become effectively more levered in stress as margin expands. Even if you never change the position, the broker changes the rules of engagement. That can force de-risking at the worst time, turning a manageable drawdown into a permanent loss of capital.

So a serious success metric in volatility trading is *drawdown under realistic constraints*. If your backtest assumes you can always hold, always roll, always rebalance, it's measuring an immortal trader. Real traders have limits: risk limits, margin limits, sleep limits.

A strategy that only survives for an immortal trader is not a strategy. It's fan fiction.

Actionable takeaway: build a "tail-aware scorecard" for every volatility strategy you test. At minimum, compute (1) annualized return, (2) max drawdown, (3) Calmar ratio, (4) worst week and worst month, and (5) time spent below the prior equity peak (time under water). Then rerun all of it under a defensive cost model (bid/ask fills and stress-widened spreads). If the strategy still looks respectable when judged by its worst moments instead of its average days, you've found something rare: an edge that can survive reality.

8.4 Building a Systematic Playbook

The market has a personality. It loves to call you at the exact wrong time.

It calls when you're up a little and feeling invincible (so you size up). It calls when you're down a little and feeling offended (so you double down). It calls when the tape is moving fast and your heart rate is trying to keep up (so you trade with your pulse instead of your rules). A systematic playbook is how you stop taking those calls.

A playbook is not a "strategy." A strategy is an idea. A playbook is an operating system: what you do, when you do it, what you *don't* do, and what automatically shuts the whole machine down when conditions turn toxic. Volatility trading punishes improvisation because volatility itself is improvisation. Your job is to be the boring part of the ecosystem.

A good playbook starts with a harsh confession: your future self is not reliable.

On calm days, you will swear you'll respect your stops. On drawdown days, you'll find reasons they don't apply. On euphoric days, you'll find reasons to press. Your brain is a defense attorney; it can argue anything. The playbook is a contract you sign while sober, designed to be enforceable while you're intoxicated on P&L.

That contract has three layers:

- *Entry rules*: the only conditions under which you are allowed to put risk on.
- *Exit rules*: the conditions under which risk comes off, whether you like it or not.
- *Circuit breakers*: rules that override both entries and exits by forcing you to stop trading, reduce size, or switch to damage control.

These layers aren't philosophy. They're trauma prevention.

The first mistake people make is writing entry rules that are emotional poetry.

"Sell premium when vol is high." High compared to what? Yesterday? Last year? The last decade? "Buy protection when the market is complacent." Complacent according to which measure: VIX level, skew, realized vol, term structure slope, credit spreads, or your gut feeling that Twitter is too cheerful?

A mechanical entry rule has to be specific enough that two different people would trigger it on the same day. If your rule can't survive being handed to a stranger, it can't survive being handed to you in a drawdown.

For volatility, that specificity usually means anchoring entries to observable market states. Examples include:

- A volatility risk premium proxy (implied volatility meaningfully above realized volatility over a defined lookback).
- Term structure conditions (contango/backwardation thresholds).
- Skew conditions (how expensive downside protection is relative to at-the-money options).
- A regime filter (risk-on vs risk-off, defined by a rule rather than a mood).

The point isn't to use all of these. The point is to stop making decisions that can be reinterpreted after the fact.

The next mistake is subtle: entry rules that ignore *tradability*.

A backtest can enter on a price that existed for a millisecond. Real execution has gravity. So an entry rule should include an execution clause: what instrument you use, what order type you allow, and what pricing assumption is acceptable.

For example: if your strategy sells an SPX option spread, your playbook might explicitly require:

- minimum open interest and volume on the strikes you'll trade,
- maximum bid/ask spread as a percentage of mid,
- a limit-order process (e.g., start at mid, then improve by x ticks, then walk away),
- a "no fill is a valid outcome" principle.

That last one is a superpower. Traders blow up partly because they treat "getting filled" as success. In volatility, refusing a bad fill is often the best trade you'll never brag about.

Entry rules also need a position sizing rule, and position sizing is where most "good strategies" become bad businesses.

Sizing in volatility can't be a simple "risk 2% per trade" slogan if your risk distribution is lopsided. Many vol strategies have a small daily variance and a huge tail. If you size off daily variance, you will oversize right into the left tail.

A practical sizing rule should reference something that expands when risk expands. Common anchors include:

- a cap on portfolio vega (sensitivity to implied volatility),
- a cap on portfolio gamma (sensitivity to underlying moves),

- a cap on margin usage (with an added buffer for margin expansion in stress),
- a cap on worst-case scenario loss under a stress test (even if the stress test is crude).

If you don't track Greeks precisely, you can still size by proxies: maximum premium collected as a percentage of capital for short options, maximum notional exposure for futures, maximum expected loss under a defined shock. The point is to size like someone who expects the market to misbehave.

Exit rules are where professionals separate themselves from hobbyists, because exits are where ego goes to negotiate.

A good exit rule doesn't try to be a mind reader. It tries to be a door.

For short-vol trades, exits often need to be *two-sided*:

- A profit-taking rule that harvests when the premium decays (because waiting for the last penny is how you get caught holding risk for free).
- A loss-cutting rule that responds quickly to regime shifts (because losses can accelerate faster than your ability to think).

The profit exit is about not being greedy. The loss exit is about not being stubborn. Both are forms of humility.

For long-vol trades, exits are emotionally inverted. The trade may bleed for weeks. The danger isn't panic-selling; it's *giving up right before the payoff* or, alternatively, refusing to take profit when the spike arrives because you want the "big one." Long-vol exits often need precommitted profit-taking rules and re-entry rules, because the payoff can be violent and brief.

What matters most is that exits are written in advance and tied to observable conditions, not vibes.

Rebalancing rules are the boring gears that keep a volatility strategy from drifting into something else.

Many strategies change their character over time if you don't rebalance. A short options portfolio can become shorter gamma as the underlying moves. A delta-hedged position can quietly become directional if you hedge lazily. A constant-maturity exposure can become a front-month gamble if you roll late.

A playbook should specify:

- how often you check risk (daily, intraday triggers, end-of-day),
- how often you rebalance (calendar-based and event-based),
- what triggers a rebalance (delta drift, vega drift, maturity drift, term structure changes),
- what you are *not allowed* to rebalance (to prevent over-trading disguised as "risk management").

That last one matters because traders love to touch things. Markets punish unnecessary touching.

If you want a single sentence to tattoo on your workflow: *Rebalancing should reduce risk, not reduce boredom.*

Now the part that keeps accounts alive: circuit breakers.

Circuit breakers are rules that override your desire to be a hero. They exist because volatility strategies can go from "fine" to "broken" in a single session. Circuit breakers don't predict the future; they respond to the present with a preplanned emergency posture.

Useful circuit breakers often trigger on:

- drawdown thresholds (e.g., if the strategy is down X% from peak, stop opening new trades for N days),
- volatility regime breaks (e.g., term structure flips into backwardation and stays there),
- liquidity stress (spreads widen beyond a threshold; fills become unreliable),
- operational errors (data outage, broker platform issues, wrong contract traded).

That last one is underrated. If you discover you traded the wrong expiration or your data feed was stale, the correct response is not "fix it and keep going." The correct response is to stop trading and audit. Most catastrophic losses start as small operational sloppiness that compounds under stress.

Circuit breakers should feel slightly embarrassing when you write them. They should feel like you're being overly cautious. Then, one day, you'll hit one and feel a wave of relief you didn't know trading could provide.

A systematic playbook also includes research discipline, because the easiest way to sabotage a live strategy is to keep "improving" it.

Once you start trading real money, the temptation to tweak is constant. A losing week makes you want to change the entry. A winning week makes you want to loosen the risk limits. Either way, you're teaching yourself that rules are optional.

So write rules for changing rules.

A clean approach is to separate *execution mode* from *research mode*. Execution mode means: you run the playbook exactly as written. Research mode means: you test changes, paper trade them, and only deploy them under precommitted criteria (e.g., after a minimum sample size or after walk-forward evidence). Without that separation, your strategy becomes a moving target. Moving targets are impossible to evaluate and easy to rationalize.

One practical safeguard: schedule strategy reviews. Not daily. Not after a bad day. Put them on a calendar (monthly or quarterly) and forbid "emergency improvements" unless something is objectively broken (like a market structure change or a product rule change). Your emotional brain wants constant iteration. Your statistical brain needs consistency.

Documentation sounds like compliance theater until you realize it's memory insurance.

Write down:

- what the strategy is allowed to trade,
- what data fields you use for signals (and at what timestamp),
- how you model fills and costs,
- what would cause you to halt trading,

- what a normal drawdown looks like and what an abnormal one looks like.

The last bullet is a psychological weapon. When you're in the middle of a drawdown, everything feels abnormal. A documented expectation anchors you. It turns panic into measurement: "Is this within the historical pain envelope, or is something new happening?"

Even if you never show this to anyone, you're building a private standard of evidence. You're choosing to be accountable to your own paperwork.

Because volatility is a social market (it's literally fear and insurance), your playbook also needs a rule for *information intake.*

On violent days, the internet becomes a siren. Everyone has a theory. Everyone is certain. Volatility spikes make certainty contagious. If your strategy is systematic, your information diet should be systematic too.

That can mean:

- a pre-trade checklist you follow before opening risk,
- a fixed set of indicators you check (not whatever is trending),
- a limit on trade frequency (because action bias is expensive),
- a rule to stop consuming news once you've executed for the day.

This isn't anti-information. It's anti-chaos. If you let the loudest voice in the room set your risk posture, your playbook becomes a mood ring.

A lot of traders resist systematic playbooks because they think rules reduce upside.

In volatility trading, rules are what allow upside to exist at all. Without them, you will inevitably oversize in the easy regime and underperform in the hard one. You'll sell the most when the premium is cheapest (because it feels safe) and buy protection only after it becomes expensive (because it feels urgent). Your P&L will become a record of your emotional timing.

Rules don't eliminate emotion. They make emotion irrelevant to execution.

And there's a deeper benefit: a systematic playbook turns trading into something you can *debug*. If performance is poor, you can isolate whether the entry logic is weak, the sizing is too aggressive, the exit is too slow, the costs are underestimated, or the regime filter is broken. If the whole thing is discretionary, you can't debug it; you can only feel bad about it.

Volatility rewards people who can diagnose, not just people who can predict.

Actionable takeaway: write a one-page playbook for a single volatility strategy you actually intend to trade. Include (1) a precise entry trigger with a timestamp, (2) a sizing rule tied to a risk-expanding measure (margin, vega/gamma proxy, or stress loss), (3) a profit exit and a loss exit written as hard conditions, (4) a rebalancing schedule that limits unnecessary trading, and (5) at least one circuit breaker that forces a cooldown after a drawdown or regime shift. Then sign it like a contract and commit to running it unchanged for a fixed review window. If you can't commit to the rules in writing, the market will happily commit you to the consequences in cash.

8.5 The Psychology of Volatility

Volatility doesn't just move prices. It moves *people.*

On a quiet day, trading feels like a spreadsheet with a heartbeat. On a volatile day, it feels like a personal referendum on your intelligence. The screen flashes, spreads widen, fills get weird, and suddenly you're not "running a strategy." You're negotiating with your own threat response. The same brain that kept your ancestors alive on the savannah is now trying to protect you from a red P&L number. It is not optimized for derivatives.

This is why volatility trading has such a distinctive psychological signature. It doesn't tempt you with one emotion; it cycles you between two: *bleeding* and *terror.*

Bleeding is the long-vol experience when nothing exciting happens.

You buy protection. You own convexity. You're doing the "responsible" thing. Then day after day the market refuses to provide drama, implied volatility leaks lower, and theta (time decay) does its slow, professional work. Your position loses a little. Then a little more. It's not catastrophic; it's worse. It's demoralizing. The losses are small enough to tolerate but frequent enough to become an identity: "I'm always wrong."

Long-vol traders rarely quit on their worst day. They quit on their 37th mildly annoying day, when they're tired of paying for insurance and start wondering if the house never burns down.

Bleeding creates a special kind of bad decision: abandoning a hedge right before it matters. It's the trading equivalent of canceling your health insurance because you felt fine for six months.

Terror is the short-vol experience when something finally does happen.

You've been collecting premium. You've been right so often it starts to feel like a law of nature. Then the market gaps down, correlations snap together, and volatility explodes upward like it was waiting for your position to get comfortable. The loss doesn't arrive politely. It arrives in chunks: first you're down, then you're down more, then you realize the bid you were relying on is now a rumor.

Short-vol traders don't panic because they're weak. They panic because their strategy contains a truth the market occasionally enforces: losses can accelerate faster than you can act. When you're short convexity, the position gets riskier as it goes against you. That's the opposite of how human intuition wants the world to work.

Terror produces the other classic bad decision: freezing. The trader watches a loss grow while waiting for a bounce that would allow an "easier" exit. Sometimes the bounce comes. The lesson becomes, "See, holding worked." Then one day the bounce doesn't come, and the account learns a harsher lesson.

What makes volatility psychology tricky is that both bleeding and terror are *rationally explainable*.

The long-vol trader can always justify cutting: "This is dead money." The short-vol trader can always justify holding: "Mean reversion." Volatility strategies are story factories. There is always a story available that turns the emotionally comfortable choice into the intellectually respectable choice.

That's why discipline isn't a moral virtue in volatility trading. It's a cognitive prosthetic.

A useful piece of psychology comes from Prospect Theory: people don't experience gains and losses symmetrically.

A loss hurts more than a same-sized gain feels good. In trading terms, this creates two distortions that show up constantly in volatility:

- *Loss aversion*: you fight to avoid realizing a loss, even when the expected value says you should exit.
- *Reference dependence*: your mind anchors on a reference point (yesterday's equity peak, entry price, "where I was this morning") and treats returning to it as a mission.

These distortions aren't just academic. They shape how traders manage options.

A short-vol trader anchored to last week's equity peak doesn't just want to survive. They want to *get back*—quickly. That urge turns sensible risk management into a revenge arc: widen the position, sell more premium, shorten the duration, pick up pennies in front of a steamroller because the steamroller is now a personal enemy.

A long-vol trader anchored to an entry price can do the opposite: hold a losing hedge far too long because selling would "lock in" the regret. The hedge becomes a moral position: "I was right to buy protection." Markets don't care about your moral positions.

Recency bias is the other villain that thrives in volatility.

Recency bias means you overweight what just happened. After a calm month, you start believing calm is normal and begin treating tail risk as a theoretical concept, like earthquakes in a city that hasn't had one lately. After a shock, you start believing shock is the new normal and begin buying protection at exactly the moment everyone else is buying it too.

Volatility markets are basically machines designed to monetize recency bias. When the world is calm, insurance gets cheaper because humans get forgetful. When the world is scary, insurance gets expensive because humans get vivid memories. If you let your beliefs track your most recent emotions, you'll tend to sell protection when it's underpriced and buy it when it's overpriced—the exact opposite of what you intended.

The cruel part is that your brain will call both decisions "risk management."

There's a behavior trap specific to options traders: confusing activity with control.

Options make you feel like you can always do something. Roll. Adjust. Hedge. Convert. Add a wing. Move strikes. Add duration. Take off the short leg. Put on a collar. The menu is endless, and the existence of a menu makes the trader feel responsible for ordering constantly.

But trading more is not the same as managing risk better.

High activity often comes from overconfidence and action bias—the deep human preference to do *something* rather than sit still. Volatility environments amplify this because movement feels like danger, and danger makes stillness feel irresponsible. The result is churn: constant tweaking that turns spreads into a slow bleed and turns a strategy into a collection of impulses.

A professional playbook doesn't eliminate adjustments. It ration them. Adjustments are treated like medicine: powerful, sometimes necessary, and harmful when taken out of anxiety.

If this all sounds grim, here's the liberating twist: the goal is not to become emotionless. It's to switch what you care about.

Volatility trading improves when you stop judging yourself by outcome and start judging yourself by process.

Outcome-based thinking sounds natural: "Did I make money today?" "Was I right?" "Did I call the move?" The problem is that volatility strategies can be right and still lose on a given day, or wrong and still win. A short-vol strategy can make money on a day you took terrible risk. A long-vol strategy can lose money on a day your hedge did exactly what it was supposed to do (it existed, it was liquid, it maintained convexity). If you reward yourself only when the outcome is green, you train yourself to prefer whatever generates immediate green, regardless of long-run expectancy.

Process-based thinking asks different questions:

- Did I follow my entry rules?
- Did I size according to my risk limits?
- Did I take exits the way I said I would?
- Did I respect my circuit breakers?
- Did I avoid making decisions while emotionally hijacked?

If you can answer "yes" on a day you lost money, that day becomes a success. That's not feel-good psychology; it's how you keep a strategy intact long enough for its edge to show up.

In volatility, the edge often arrives in bursts. You need to still be in the game when the burst comes.

One of the best psychological tools in volatility trading is a pre-commitment ritual that feels almost silly until you try it.

Before you enter a trade, write down two numbers: the amount you expect to lose if the world stays boring, and the amount you could lose if the world stops being boring. Write them as dollars, not percentages. Your brain can ignore percentages. It can't ignore rent money.

Then write down the exit condition that would force you out. Not what you *hope* you'll do. What you will do. This turns a vague intention into a decision already made. When the market later tries to negotiate with you, you can point to a decision your calmer self already signed.

This matters especially for short-vol trades because the biggest losses often arrive alongside a persuasive internal voice: "Just wait for mean reversion." Sometimes mean reversion shows up. Sometimes it doesn't. A pre-committed exit isn't about being right; it's about not letting one trade define your career.

Long-vol traders need a different pre-commitment: how to survive the boredom without sabotaging themselves.

A common failure mode is "hedge shaming": feeling stupid for paying for protection in calm markets. The trader cuts the hedge, or reduces it, or starts selling premium against it in a way that quietly cancels the convexity. The hedge becomes a decorative object rather than a functional one.

A better approach is to explicitly budget for bleed, like an insurance premium you are willing to pay. If you can't tolerate that premium, you can't afford the hedge—not financially, but psychologically. And if you can't afford it psychologically, you'll bail right before it pays.

There is no virtue in owning long volatility if you only own it during crises. That's like buying an umbrella after you're already wet.

Short-vol traders need a complementary discipline: learning to treat "easy money" as a danger signal.

When selling premium feels effortless, your risk is often quietly rising. Calm markets compress implied volatility, which makes options cheaper, which tempts you to sell more contracts to hit the same income target. This is the trap: you increase size as compensation for low premium, effectively levering up in the very environment where the strategy's payout is smallest and the tail risk is building.

This is why position sizing rules in volatility trading are psychological rules. They prevent you from seeking emotional satisfaction (steady income) by increasing structural fragility. They force you to accept smaller gains in calm periods rather than turning calm into a leverage invitation.

A short-vol trader who can sit through a low-premium regime without forcing trades has something rare: emotional liquidity.

There's also the identity trap: confusing your self-worth with your model.

Volatility attracts smart people. Smart people like being right. Options provide endless ways to build narratives about being right: "I understand skew." "I understand carry." "I understand regime shifts." When the market disagrees, it can feel like an attack on competence rather than a normal part of probabilistic outcomes.

A trader who needs the market to validate their intelligence will take trades that satisfy ego rather than expectancy. They'll hold losers to avoid admitting error. They'll press winners to prove the point. They'll ignore new information because it threatens the story.

The antidote is to treat trading as operating a machine, not expressing a worldview. Your job is not to be right. Your job is to follow a process that has positive expectancy and bounded damage. The market can call you wrong every day and you can still succeed if your risk is structured correctly.

Volatility trading becomes calmer when you stop demanding respect from the tape.

A practical psychological technique that works far better than it sounds: write a "panic protocol."

When volatility spikes, your cognitive bandwidth shrinks. You will forget your own rules. You will see only the most salient number (the loss) and lose the ability to compare it to your plan. So create a one-page protocol for high-stress moments. It can be as simple as:

- No new trades for 30 minutes after a large move.
- Check position exposures and margin first, news second.
- Execute exits according to prewritten rules; no discretionary overrides.
- If spreads exceed a threshold, switch to reduce-risk orders only.
- After action, step away for a defined period.

This isn't about being timid. It's about preventing the single most expensive behavior in volatility trading: improvisation under adrenaline.

Actionable takeaway: choose whether you are structurally "bleeding" (long vol) or structurally vulnerable to "terror" (short vol), and then write one pre-commitment rule that directly targets your personal failure mode. If you're long vol, set a monthly "bleed budget" and a rule that forbids cutting hedges impulsively during calm markets. If you're short vol, set a maximum risk exposure (margin or stress loss) that cannot be increased when implied volatility is low. Then add a simple panic protocol for spike days. The goal isn't to feel fearless; it's to make sure your fear can't touch the controls.

8.6 Practicalities: Taxes and Compliance

The fastest way to ruin a perfectly good trading edge is to ignore the unsexy stuff.

Taxes and compliance don't feel like "trading." They feel like paperwork you'll deal with later, after you've made the money. The market loves that attitude, because it turns profitable strategies into expensive hobbies. In volatility trading, the details matter even more because you're often trading products that live in special tax categories, and because it's easy to talk about performance in ways that are accidentally misleading (to others or to yourself).

Think of this as operational risk: not the risk that the VIX spikes—but the risk that April 15th spikes.

Start with the headline most U.S.-based volatility traders should know cold: *Section 1256*.

Certain derivatives are treated as "Section 1256 contracts" under U.S. tax rules. The practical punchline is twofold:

- *Mark-to-market*: if you hold these contracts at year-end, the IRS generally treats them as if they were sold at fair market value on the last business day of the year, even if you didn't close them. (irs.gov)
- *60/40 treatment*: gains and losses are generally split 60% long-term and 40% short-term, regardless of how long you actually held the position. (irs.gov)

For a lot of volatility traders, that 60/40 split is the hidden "alpha" nobody puts in the backtest. If you're actively trading something that qualifies, you can end up with a meaningfully lower blended tax rate than you'd get from pure short-term capital gains. That doesn't make a bad strategy good, but it can make a good strategy *survive* after taxes.

The first practical lesson: you can't evaluate a volatility strategy purely pre-tax if your instruments fall on different sides of the tax fence. Two strategies with identical pre-tax returns can have very different after-tax lives.

The second practical lesson: you have to know what you're actually trading.

Volatility traders casually say "index options" as if they're all the same. They aren't. Some products are treated as non-equity options (often associated with broad-based indexes), some are equity options, some are futures, and some are ETFs that merely *hold* derivatives. The same economic idea ("short volatility") can be implemented through instruments with very different tax outcomes.

That's why a professional habit is to tie your strategy definition to the instrument type in your broker statements, not to the story in your head. Your strategy might be "short vol," but the IRS sees "options on securities" or "regulated futures contracts" or "a fund that issues a 1099." Different world, different rules.

If you're not sure whether something is Section 1256, don't guess. Get it confirmed by your broker's tax reporting, your 1099, contract specs, or a tax professional. Guessing here doesn't make you bold; it makes you randomly exposed.

Now for the part that surprises people the first time they run into it: *mark-to-market means you can owe taxes without selling.*

If you have open Section 1256 positions at year-end that are up, you may recognize a gain because the position is treated as sold (and then, effectively, reacquired) at year-end fair value. (support.taxslayer.com)

That can be great for cleaning up accounting and avoiding some of the gamesmanship of realizing gains. It can also be annoying if you're running a strategy that holds exposure across year-end and you're not managing cash for taxes. Many traders discover this the hard way: January feels like a fresh start until you remember you prepaid for last year's profit.

A practical rule: if your strategy uses 1256 products and you're having a strong year, treat tax cash management as a required position. Don't wait until the forms arrive to discover you spent the IRS's portion on "reinvesting."

Where does this get reported?

Form 6781 is the form that covers gains and losses from Section 1256 contracts and also gains/losses under the straddle rules (Section 1092). (irs.gov)

That sounds like trivia until you realize what it implies operationally: if you trade these instruments, your taxes are not "just Schedule D" in the simple sense. Even if your broker provides summaries, you still want to understand what bucket your trading falls into and why. The trader who can't read their own tax reporting is flying IFR with no instruments.

(And yes, this is where a good CPA pays for themselves if you're active and trading multiple instruments. The goal isn't to find loopholes. It's to avoid accidental landmines.)

Next: wash sales, the rule that turns a normal-looking loss into a delayed tax benefit.

Most traders know the headline version: sell a stock at a loss and buy it back within a window, and the loss may be disallowed (or, more precisely, deferred and rolled into basis). What volatility traders often miss is that wash sale rules can apply not only to stocks, but to *contracts and options to acquire or sell stock or securities.* (irs.gov)

And there's a second detail that matters for vol traders who wander between asset classes: the wash sale rules *do not* apply to losses from sales or trades of commodity futures contracts and foreign currencies (as described in IRS guidance). (irs.gov)

This creates a common real-world confusion:

- You trade options on an equity ETF or single stock and assume "it's options, wash sale won't matter." It can.
- You trade regulated futures and assume wash sale applies the same way it does to stocks. It generally doesn't.

Even if you're not trying to "tax trade," wash sales can happen by accident when you roll positions, adjust hedges, or run a systematic strategy that re-enters quickly after a stop-out. If you think you harvested a loss for tax purposes and your actual tax reporting says otherwise, you'll learn the definition of "substantially identical" in a very personal way.

The trading implication is simple: if your strategy depends on frequent stop-outs and re-entries in securities/options, build tax awareness into the execution calendar. A rule like "cooldown before re-entry" can reduce behavioral overtrading *and* reduce wash sale complexity. Sometimes discipline pays twice.

Then there are *straddles*, which are less about one trade and more about how the IRS interprets your portfolio.

A straddle, in tax terms, generally involves *offsetting positions* with respect to personal property. (law.cornell.edu) In human terms: you own positions that substantially hedge each other. That can be as obvious as a long call and long put on the same underlying, or as subtle as a portfolio where one position's gain is designed to offset another's loss.

Why does this matter? Because straddle rules can limit or defer losses. Under Section 1092, losses on one leg may only be taken into account to the extent they exceed unrecognized gains on the offsetting positions (with deferred losses generally carried forward). (law.cornell.edu)

This is the part that breaks traders' intuition: you can be down on one leg, up on the other, and still not get the tax treatment you expected on the losing leg when you expected it. The system is trying to prevent a simple game: realizing losses now while keeping offsetting gains unrealized.

For volatility traders, straddle treatment can sneak in through common structures:

- option spreads and collars,

- delta-hedged option positions,
- hedged portfolios that intentionally offset risk exposures.

None of this means "don't hedge." It means: when you build offsetting positions, don't assume tax outcomes match P&L outcomes on your preferred timeline. If your strategy's edge is thin, timing differences can matter.

Also note the practical connection: Form 6781 explicitly references reporting for Section 1256 mark-to-market and for Section 1092 straddles. (irs.gov) If you're active in these instruments, the paperwork is pointing at the complexity for a reason.

Now pivot from taxes to compliance, which for many retail traders sounds like something that only matters if you're running a fund or selling a course.

Here's the real framing: compliance is how you avoid lying with numbers.

Even if you never manage money for others, you will market to *someone* eventually: a spouse, a business partner, a future investor, or yourself six months from now when you're deciding whether to size up. The temptation to present performance in the most flattering way is universal. Volatility strategies, because of their skew and tail risk, are especially easy to misrepresent with innocent-seeming choices: gross returns instead of net, mid marks instead of executable marks, cherry-picked regimes, or backtests shown without the ugly footnotes.

Regulators care about this because misleading performance presentation harms investors. But you should care because misleading performance presentation harms *you*. If you build confidence on inflated numbers, you will size based on fantasy and then wonder why reality feels unfair.

If you are (or may become) an investment adviser presenting performance, the SEC's marketing rule (Rule 206(4)-1 under the Advisers Act) provides a useful template for what "serious" looks like.

It places conditions on the use of hypothetical performance, including requirements around having policies and procedures designed to ensure relevance to the intended audience and providing sufficient information about criteria, assumptions, risks, and limitations. (law.cornell.edu) It also defines net performance as performance after deduction of all fees and expenses a client would pay, with specific guidance around model fees. (law.cornell.edu)

Even if you're not an adviser, that mindset is worth borrowing: if you show a backtest or model, you should be able to answer, crisply:

- What assumptions did I use (fills, spreads, fees, slippage)?
- What data was available at the time (no lookahead)?
- What risks could make this fail (tail events, liquidity)?
- What would the result look like net of realistic costs?

If you can't answer those, the performance isn't "wrong" so much as *unsubstantiated*. And unsubstantiated confidence is how traders blow through their own risk limits.

If you operate in commodities/derivatives promotion contexts (for example, as a CTA/CPO, or producing promotional material under those umbrellas), the CFTC and NFA have their own well-known focus: hypothetical and simulated performance.

CFTC Rule 4.41 is often associated with requiring prominent cautionary language for simulated/hypothetical performance results; CFTC Letter No. 01-60 discusses that presenting hypothetical performance in commodity interest contexts triggers these requirements and that prescribed cautionary statements must be displayed. (cftc.gov) The key point isn't memorizing the exact disclaimer text from memory; it's understanding the principle: hypothetical results can be dangerously persuasive, so the disclosure obligations are strict.

For a personal trading playbook, the analogous rule is simple: whenever you look at a backtest, force your eyes to see the limitations with the same prominence as the equity curve. If you bury the caveats, you'll eventually trade as if the caveats don't exist.

Compliance also includes broker realities: options approval, margin permissions, and recordkeeping.

Even a perfectly designed volatility strategy can become unusable if your broker doesn't approve the required option level or if margin treatment makes the strategy too capital-inefficient. This isn't an academic nuisance; it's an execution constraint that changes optimal strategy design. A spread that looks elegant on paper might require less buying power than a naked position, but might also limit flexibility in stress. Your playbook should treat broker permissions as part of the system, not an afterthought.

Recordkeeping is equally unglamorous and equally decisive. Volatility strategies often involve many legs, rolls, and adjustments. Without meticulous records, you can't:

- reconcile broker statements to your strategy logic,
- diagnose performance leaks (slippage, spreads, roll timing),
- substantiate results if you ever present them,
- make tax time a process instead of a panic.

A useful habit is to store a "trade journal" that is more like an audit log than a diary: timestamps, instruments, size, rationale (signal state), expected risk, actual fill, and the reason for exit. If you can't reconstruct what you did and why, you can't improve it—and you can't credibly explain it.

Actionable takeaway: treat taxes and compliance as part of strategy design, not cleanup. For every volatility strategy you trade, write a one-paragraph "tax and reporting note" that answers: (1) is the primary instrument likely a Section 1256 contract with mark-to-market and 60/40 treatment, (2) could wash sales apply based on re-entry behavior, and (3) could the position set be considered offsetting (straddle-like), affecting loss timing. Then build an operations checklist: download year-end statements early, reconcile trades to your own logs monthly (not annually), and if you ever share performance with anyone, present it net of realistic costs and alongside plain-language limitations. The goal isn't to be paranoid; it's to avoid paying tuition for mistakes that were avoidable with one afternoon of discipline.

www.ingramcontent.com/pod-product-compliance
Lightning Source LLC
LaVergne TN
LVHW052020200926
848481LV00005B/1536